ROUTLEDGE LIBRARY EDITIONS:
HUMAN RESOURCE MANAGEMENT

Volume 23

SOCIAL STRUCTURE AND PERSONALITY IN THE FACTORY

SOCIAL STRUCTURE AND PERSONALITY IN THE FACTORY

PAUL LAFITTE

Routledge
Taylor & Francis Group
LONDON AND NEW YORK

First published in 1958 by Routledge & Kegan Paul Limited

This edition first published in 2017
by Routledge
2 Park Square, Milton Park, Abingdon, Oxon OX14 4RN

and by Routledge
605 Third Avenue, New York, NY 10017

Routledge is an imprint of the Taylor & Francis Group, an informa business

British Library Cataloguing in Publication Data
A catalogue record for this book is available from the British Library

ISBN: 978-1-138-80870-6 (Set)
ISBN: 978-1-315-18006-9 (Set) (ebk)
ISBN: 978-1-138-28584-2 (Volume 23) (hbk)
ISBN: 978-1-315-26876-7 (Volume 23) (ebk)

Publisher's Note
The publisher has gone to great lengths to ensure the quality of this reprint but points out that some imperfections in the original copies may be apparent.

Disclaimer
The publisher has made every effort to trace copyright holders and would welcome correspondence from those they have been unable to trace.

SOCIAL STRUCTURE AND PERSONALITY IN THE FACTORY

by
PAUL LAFITTE

LONDON
ROUTLEDGE & KEGAN PAUL

First published in 1958
by Routledge & Kegan Paul Limited
Broadway House, 68–74 Carter Lane
London E.C.4
Printed in Great Britain
by Butler & Tanner Limited
Frome and London

PREFACE

It may be as well to make clear both the nature and the origin of the work that is reported in this book.

Firstly, this book follows the two previous studies of social behaviour that were made in this Department (*Social Structure and Personality in a City* and *Social Structure and Personality in a Rural Community*) in the general sense that it too is a study of social behaviour. It is much more limited than the first two studies, as it is concentrated solely on the behaviour of industrial workers; but it is on the same general lines, in that it is a study of the worker both in his general social context and in his local situation, and not merely a differential study of abstract qualities of personality.

The book derives from the interest of the Espada Educational Trust in industrial tensions generally. This Trust, which was set up in Melbourne by the late Mr. E. C. Dyason, has as its general aim to promote the study of tensions between groups of persons; and it has been active in the support of a number of projects within its general terms of reference. The Trust took the broad view that industrial tensions would best be studied by a general investigation of the work situation from the worker's point of view (which is very little known), and by the correlation of this point of view both with the local situation (as the observer sees it) and with the management's point of view. In consequence, the investigation that is reported had wide terms of reference, and the report makes particular reference to industrial tensions only in so far as these could be related to specific circumstances. The emphasis throughout, that is to say, is on the worker's general orientation to his situation and not specially on his discontents.

Most of the work reported was carried out in factories in Melbourne in 1951 and 1952, but the account includes some notice of an investigation made in 1950, the only one that was made primarily at the request of the factory management. Except only for this section, the considerable financial burden of the project was carried for the most part by the Espada Educational Trust and to a lesser extent by the General Research Fund of the University of Melbourne. Without the ample, not to say generous, backing of both these sources, the project would have been impossible.

While the responsibility for the field-work, for the analysis of data and for the preparation of the final account was primarily mine, I must acknowledge my deep indebtedness to Professor O. A. Oeser, both for his general guidance and support throughout the project and for his criticisms of the first draft. At the same time, I must acknowledge with gratitude the labours of my colleagues who had various parts in the project: Mrs. Lynley Graham, Mrs. Cecily Katz, Mr. F. M. Katz, Mr. P. A. Loftus, Mr. R. E. C. Penny and Dr. Wilma Scott (who conducted a large part of the interviews); Dr. Drury Oeser, Mr. L. H. Rosenberg and Mrs. Noel Shaw (who were research assistants); and lastly Mrs. Irene Kinsman, whose assistance in the editing of both the interview records and the final manuscript was invaluable.

In view of the list of acknowledgements, it may seem somewhat remarkable that the final report is published years after the conclusion of field-work, for no less than twelve university graduates were concerned in the project. In fact, only three of these were engaged in it as their sole duty, and only four (whether full time or part time) were engaged in it for more than one year. Further, for a year and a half it was impossible to engage any research assistants on the terms that could be offered. It may be bad form to draw attention to these incidental difficulties of staffing, and it must be made clear that no criticism whatever is implied of the bodies which financed the project: on the contrary, it is clear that they supported it very generously, within the limits of their resources. It is high time, however, that someone made the obvious point. Until the social sciences can occasionally spend on staff sums of the order that physical scientists and technologists spend as a matter of course on equipment, research in the social sciences is bound to be lengthy and laborious and (possibly) out of date before it is published.

My acknowledgements are due especially to all the persons who took part in the investigation, whether managers, foremen or workers. That all of these persons must remain anonymous is obvious: I can only thank them for their assistance in the general advancement of knowledge.

Two comments on the organization and layout of the book may help the reader. Firstly, it seems to be usual nowadays that texts in psychology or sociology refer to everything that has ever been written on the subject or on remotely related subjects. This serves both to display the author's erudition and to baffle and irritate the reader; so that it serves no useful purpose. It is especially irrelevant in an empirical account of a particular investigation. I therefore made no attempt, at any stage, at a comprehensive survey of the whole literature of industrial psychology or sociology; and I have deliberately limited the references given to the few that are directly relevant.

The second point is perhaps less obvious. Throughout the book, but especially in chapters II and XI, considerable detail is given of field procedures and of the methods of analysis used. In one sense, this detail is necessary. With an investigation that is somewhat unusual in aim and unconventional in the methods of analysis, the matters detailed cannot be taken for granted altogether. One need not even refer to recent discussions of methodology in the social sciences to make the obvious point that conclusions depend very much on what was done. On the other hand, for many purposes the reader might prefer to take methodology for granted, in both aspects, and would then only be held up by the longer sections on this. There seemed to be no simple way of avoiding both difficulties, but I decided that the least unfair compromise was to leave in the methodological sections and to note (in chapters II and XI) where these could be skipped by the reader who does not require them.

PAUL LAFITTE

Department of Psychology
University of Melbourne
October 1957

CONTENTS

CONTENTS

LIST OF TABLES

CHAPTER I

INTRODUCTION

THE account of an extended investigation of behaviour at work starts appropriately from the consideration of what is known in this field and what general views have been put forward. It is a curious feature of empirical psychology that little enquiry has been made into behaviour in ordinary real life situations. Besides this, most of the investigations in the industrial field have been of particular problems requiring immediate attention; and work of this kind often depends little on the general body of psychological knowledge or even on common experience. In consequence, findings are few and scattered and not always clearly related to the more numerous general explanations of industrial behaviour. However, the variety of both the general analytic approaches and the empirical findings can be summarized under four main themes.

Firstly, the person works because he must but also because he finds some relation between his task and its concomitant conditions, on the one hand, and his personal goals and values, on the other. This is roughly the view put by Mace (1948), and implies that nobody would work unless he found some satisfactions in it. The second view rather contradicts this: industrial behaviour is due to the person's place in society and in particular to his relationship to the means of production. Sharp's argument (ch. XVII in Oeser and Hammond, 1954), to this effect would thus tend to the conclusion that work satisfies the employer because he is an employer but cannot satisfy the worker because he is only a worker. Another view which has become increasingly popular in recent years (cf. Gardner and Moore (1950) or Miller and Form (1951) as examples) is that industrial behaviour depends on the person's social relations at work. It is these, which may be incidental to his task, which satisfy or dissatisfy him. Lastly, there is the view of what might be called the classical phase of industrial psychology, as set out by Scott (1950) or Viteles (1954), among many others. This view derives directly from the earliest studies of the worker as a physiological mechanism, but is now much more complex. It now tends to ascribe industrial behaviour to a large number of small and apparently unrelated factors: physiological, environmental, sociological and personal.

It is not suggested that this very bare summary does justice to the views or to the authors mentioned, nor, indeed, that it represents all the views that have been put forward on industrial behaviour. What it does do is to point to the first necessary definition, that of industrial behaviour, or behaviour at work, in general. If this is taken to be merely what men do because they must, compared with what they do because they choose, then there is hardly a question for psychology. It is a fact of social organization that men must work and may play; and the only comparative question for psychology here would be why a few men disregard these rules when most observe them. If the barest definition of behaviour at work is extended to include the wide variety of specific behaviours that is required by the variety of tasks that men are paid to do, psychology still has almost nothing to say. It can only point to the self-evident fact that men doing different tasks behave differently because they have been trained to do different tasks. Comparative questions about the differences between tasks belong to economics and technology and to the history of industry.

However, it is sufficiently evident in common experience that the behaviour of men at work is not simply determined by the actual task. A man may work efficiently or inefficiently, cheerfully and co-operatively or resentfully and obstructively, and so on. These differences may be found between men doing what is apparently the same task, so that the least question for psychology is the study of the actual range of variety in behaviour in a specified task. But differences between men doing different tasks are equally marked, and these too are within the scope of psychology. Finally, so are differences between men in their covert behaviour, or in their opinions of their tasks. It is true that differences in behaviour have been insufficiently studied, but it is likely enough that they do depend to some extent and in some sense on task differences, social relations, the workers' place in society and his personal goals. Thus far, it can be said that each of the four views summarized above is to some extent true.

But as it stands this means only that the study of behaviour at work must take into account a great many factors. It is far removed from a coherent plan of investigation which sets out the necessary comparative questions in order. Making a plan requires that the person's behaviour at work should be considered in detail and in serial order. The person, that is to say, must be considered historically in principle, even if only over a short time. What the person does at work he does only after more or less thought or deliberation; then he decides, then he acts. However brief and however unreportable it may be, there is always some series of covert behaviour preceding action; and it cannot be ignored for it is the effective cause of action, and may therefore be the cause of different actions, in so far as it may differ from one man to another. But before the person can evaluate his work situation, he must notice or perceive it; and, before the psychologist can attempt to reconstruct situations as they are perceived by different persons, he must know the differences between situations as they appear to him, a detached and presumably expert observer.

In practice, the specification of the series of behaviours at work is implied by generalizing from strictly historical series over short periods to series of recurrent behaviour. It is assumed that the series of the strictly defined single event can be disregarded. The person's situation at work, from the point of view of the expert observer, was such and such on last Tuesday at eleven in the morning, but it was not exactly the same an hour before, let alone a day before or a week before. The person's perception of his situation at these four times was different, and so was his consequent decision on what to do, and his eventual actions. The differences throughout the series might be specifiable if there were time enough to study them, but this is unlikely. The author has argued elsewhere (1957, pp. 89–92) that psychology is obliged to condense the time-scale of the events it studies. In consequence, a general industrial investigation cannot be concerned with daily or weekly variations of behaviour, it must concentrate on recurrent long-term behaviour. One must assume that the person has been in the situation for some time, so that the way he perceives it is recurrent, if inarticulate; or would have been the same as his present articulate evaluation if he had been asked to make an evaluation of this kind three or six months ago. Finally, one must also assume that his behaviour in the work situation is much the same now as it was three or six months ago; that is, that it is settled recurrent behaviour. These are considerable assumptions, but they must be made if any general investigation is to be planned. It will suffice here to note that these assumptions are not necessary in the light either of common experience or of psychological theory, they are necessary only in practice for the particular purposes of psychological enquiry; their general validity is one of the intriguing but barely attempted questions of psychology.

THE SERIES OF BEHAVIOURS AT WORK

The first of this series is the person's situation as it is seen by a detached observer. This includes at least six elements, or partly overlapping situations, which may be taken in order:

1. The person's actual task. From the point of view of the observer, this is the cycle of operations that he must perform in

order to satisfy his employers, or his customers, if he is self-employed. It is a bare minimum specification of the actions that must be performed before the observer can say that two men at any occupational level are carrying out the same task, similar tasks or widely different tasks. Even though it is a bare minimum specification, there is an enormous variety of tasks that may have to be taken into account.

2. The system under which the person is paid, which includes several variables. He may be paid the bare minimum which has been defined by law or agreed, or above the minimum. Beyond this, he may or may not be able to work overtime and may or may not be paid various bonuses (e.g. for punctuality, or on the basis of annual profits) and may or may not get various allowances (e.g. tool money, dirt money, standby money). Beyond this, again, he may be paid straight day rates or piece rates (the latter, it should be noted, hardly apply to Australian male factory workers). Finally, he may be paid a weekly wage, with or without the above variations, or a monthly salary, usually with no variations. In this respect, there is less variety to be expected than with the tasks alone, but there is still enough.

3. The persons who direct the worker's activities (his bosses, of various grades) may be expected to vary. The variety here is less easily specified in general terms because it has been insufficiently studied and is less obvious. However, it might be said that the person's immediate boss may be either person-centred or production-centred; and this crude dichotomy is perhaps not altogether unlikely for the foremen of factory workers. Besides this, the top management under which the person works may be autocratic or co-operative, benevolent or restrictive; and many other classifications are possible. Whatever hypothetical classifications are proposed, there will therefore be considerable variety in the generalized (or recurrent, or habitual) behaviour of the person's bosses, as this behaviour is seen by the detached observer for all grades of boss taken together.

4. The other persons with whom the person works may be expected to vary. The variations in the number of persons with whom he works, and in his specific working relations with them, are included in the account of variations in the task. Beyond these, however, there will be variations in the psychological structure of the work group. This may be co-operative throughout, co-operative only in parts or unco-operative throughout. The corresponding structural (or sociometric) descriptions would be: mutual dependence (or choice), some mutual dependence and some isolates, isolates only or each man for himself. This specification by no means exhausts the possibilities of variety in the person's

relations with his workmates, as seen by the observer; but, even so, the possible varieties are considerable.

5. It may be necessary to consider the factory as a social organization. The person's present and recurrent relations with his bosses have been sufficiently specified but beyond these are his prospective relations. Both particular factories and whole industries differ in the extent to which they can offer the worker of any grade secure and continued employment and prospects of promotion. In these respects at the least it will be necessary to differentiate between work situations; and, depending on the importance of these and other possible variations, it may be necessary to break up this set of variations into a number of separately specified elements.

6. The person's engagements in the work situation, then, are likely to be highly complex. But these are by no means his only engagements. Besides the activities which are directly required by the work situation or which are subsidiary to it, the person also does things as a citizen or as a member of a particular country, as a member of a social or occupational class, as a member of a particular kind of family group (in the limiting case, himself alone) and as a member of various friendship groups; and he also does some things by and for himself alone. It is, in principle, ridiculous to regard this wide-ranging collection of conceptual and personal behaviours as being together no more than the sixth and last of the person's 'situations' as described by the detached observer. In practice, however, it is not only possible but reasonable, on two grounds. Firstly, comparative questions can be asked about the person's behaviour as a citizen only if the investigation includes persons in more than one society. As this is unlikely, and is not the present case, it can be disregarded. The second ground of restriction is less obvious, as it derives from the observer's relation to the persons observed. Being an expert, he is capable of observing behaviour over a wide range; but, being a person who is immersed in history like his subjects, what he can actually observe is much more limited:

(i) the person's present occupational class and family responsibilities can be taken from his own statement. The former is an obvious public fact and the latter, if not so obvious, is hardly open to mistake or deliberate distortion; the person is married, with three dependent children or he is single, and these are the facts. But these are the only facts which the observer can safely assume merely from the person's report (or from the factory records). What the person does in political meetings or when he votes in trade union affairs in the factory or in general union meetings, are all matters that he may be willing to report, under conditions. In the present state of knowledge, it would be most unsafe to assume that what he reports is what he does; and it is not practically possible for an observer, however 'participant' he may be, to see what the person does in these affairs. The same holds, more obviously, for the person's family behaviour. Beyond the bald facts of the situation, he may be willing to report how he gets on with his wife and children; but his report obviously cannot be checked by systematic observation and therefore cannot, in the present state of knowledge, be accepted on the same level as the report of his bare sociological responsibilities.

(ii) the same difficulty applies even more forcibly to the description of the person's engagements with his friends (group or individual) and with himself. In both cases, all the observer has to go on is the person's own report, for neither field of activity is, in its nature, open to observation. In both cases, there is quite insufficient general knowledge of the field of activity for the observer to be able to form a definite conclusion about the person's actual engagements.

The practical conclusion, then, is that the specification of the person's engagements outside the work situation must be limited to a statement of his present occupational class and present family responsibilities (or life phase). Whether or not these bare sociological facts have any relation to the person's covert evaluative behaviour or to his overt behaviour at work, they are the only facts that the observer can rely on as sufficiently evident or sufficiently established from the person's report. Being bare sociological facts, they are sufficiently discriminated from the other elements in the person's whole situation by their grouping together as the sixth element.

The second part of the series of behaviours at work is the person's covert behaviour, or what goes on inside him between his being stimulated by the work situation (or other situations, or a combination of situations) and his acting. Unfortunately, the observation of Krech and Crutchfield (1948, p. 80) is still true, that little is known about how the person perceives the situations in which he acts. However, it is this information which is most necessary. There is no reason to suppose that the person at work sees his situation in the same way as the observer does (where that applies), or that all persons at work see in the same way situations that are identical from the observer's point of view. On the contrary: the very reason for the existence of empirical psychology is that behaviour is not simply bound to the situation, but is complexly derived from it, through the person's various perceptions and intentions. Even though little is known, a general and hypothetical statement of the matter to be investigated is therefore necessary.

7. The person's perception of his situation at work is clearly relevant. One must know how he says he does his task, which may or may not be the same as what he actually does (as observed) or what he is supposed to do. Equally, one must know how the pay system works and what his bosses and mates do, according to him; which again may or may not be the same as the observer's report on the pay system or the behaviour of specified bosses and work groups. Lastly, one should in principle include the person's perception of the place of work as a social organization, though this may be difficult to elucidate in practice, as in many cases an articulate report on this will be beyond the person's level of verbal sophistication.

8. The next necessary element is the person's perception of his various engagements outside the work situation. His reported class behaviour (party or trade union affiliations and activities) and his general opinions on social classes and his own place in society must be included here. So must his general report of how his family is organized (if he is married): what is the division of responsibilities, what are the rules or practices for bringing up the children. This leads on to the person's report of his unprescribed activities, or what he does because he chooses to, not because he must. This by itself includes a considerable range of activity: membership in clubs and societies, domestic activities, sport, family recreations, passive entertainments, and so on. Lastly, one must include the person's report of his projects or aspirations for the future. These again may range considerably, from projects at present in hand (e.g. building a house, or improving his job qualifications by study), through to vaguely articulated hopes for the unstructured future (e.g. win a lottery prize or retire).

9. The reason for specifying the person's perception of these engagements as necessary information is that he may perceive himself in the work situation in relation to his other engagements. The work situation may or may not be relevant, as he perceives it, to his ideas of his proper place in society, or of the importance of his family responsibilities, his unprescribed activities or his aspirations for the future. Whether or not he has any definable goals of central importance, the possibility must be explored that his behaviour at work derives in part from his reported or inferred relative evaluation of his engagements outside work.

10. The list is still not complete. Even if the person has stated what he does at work or what is done to him, followed by what he does elsewhere or what he thinks of his place in society, followed again by what relations he sees between his activities at work and elsewhere, one still does not know his final evaluation or practical judgement of the work situation. This may be for the most part

given in the person's description of the work situation as he sees it. It may be expected that his report of what he does in his task, of how the pay system works and of how his bosses and mates behave, will include many evaluative comments from which one can infer his overall practical judgements. On the other hand, there may be discrepancies between these incidental evaluative comments and his final judgements. As the latter are the last step in order before action, it is necessary to ask the person specifically how he likes the four or five distinct aspects of his work situation and, in particular, what are the good and bad features of each.

The person's situation, as seen by an observer and as seen by himself, have been considered in turn. The way is therefore clear for the definition of the third main part of the series of behaviours at work, which is what the person actually does. His behaviour at work derives from his final evaluation of the various aspects of the work situation, and this evaluation derives in turn from the way he perceives the work situation and his other engagements; and these perceptions derive, in part at least, from the actual situations, as described by the observer. But it cannot be assumed that even the last link, let along the whole causal series, can be rigidly and uniquely defined. In any case, it is possible that action in the work situation is not simply and uniformly derived from the person's final evaluation of this situation because the possibilities of action are to some extent open, so that the decision between different lines of action can be determined only by trial or experiment. This is the usual case in any complex field of ordinary daily life behaviour (cf. the author's discussion of this (1957, pp. 84–9)). In the present state of ignorance of the second or intervening stage of differential perception and evaluation, it is more than likely that there will be some relation, but by no means a uniform relation, between the person's overall evaluation of his work situation and his actions. The person's behaviour at work, then, must be studied in its own right. Here too, unfortunately, there is not enough information to support the statement of a detailed scheme of investigation. There have been some studies of productivity, in situations in which that is definable for the individual worker. There have been more studies of marginal or withdrawing behaviour, such as absenteeism, sickness and accidents and labour turnover. But very little attention has been paid to what the person actually does in the work situation, as distinct from what he is expected or assumed to do. In consequence, in this case too only a general and hypothetical statement of the field to be investigated is possible; and this requires a brief review of what he can do. The person's behaviour in the work situation depends to some extent on the expectations of the persons who have effective control or direction of his job. His actions in carrying out his task are usually prescribed by rule and are relatively closely supervised. His relations with his immediate boss are prescribed by rule only in general terms, and thereafter by local custom. His relations with his workmates are prescribed, if at all, only by local custom, except insofar as they are necessarily involved in the performance of the task. In principle, variations in all respects should be considered. In a general investigation of behaviour, which is not specially concerned with management's problem of productivity, what matters is simply what the person at work does. There are three general possibilities:

11. The simplest possibility is that the person may do what is expected of him, in all respects, by the persons who control his work. He may carry out his task exactly as it is laid down by his boss, he may behave towards his boss exactly as the customary boss-worker relation requires, and he may act together with his workmates only and insofar as this is necessary for the execution of the task. In this somewhat unlikely case, the factory worker at least would probably find secondary complications in his social relations. However that may be, the general possibility must be allowed for, that the person at work simply does what is expected of him as a worker.

12. Another possibility is that the person may do what his mates do. As described in the pioneer work of Roethlisberger and Dickson (1939, chs. XVII–XXIII) this may involve only the observation of the local group's rules with regard to the restriction of output. But this implies that the person does what is officially expected of him, for the most part or to some extent; and also does other things which are at least additional to, and at most contradictory of, what he is officially expected to do. Besides the general example of restriction of output, there are many customary practices that may depend on the definition of the working group. Some examples are : unofficial breaks from work, unauthorized short cuts in job processes, the customary but not official allocation of jobs in the work group, and the custom of the group's concealing or obliterating individual mistakes. It must be assumed that any worker will meet the expectations in productivity of the persons who control his work to the least extent which is acceptable. It cannot be assumed, however, that going beyond this minimum is his first or only goal in the work situation: a compromise between this and other group goals is at the least a possibility that must be allowed for.

13. The last possibility that requires consideration is that the person does what is expected of him to some extent (as he must, as he works for a living), but as far as he can does what he judges

to be appropriate to his personal goals and values. Conventionally, he goes through the motions as far as is necessary but puts his real effort into what concerns him as he sees it. In this case, the classical case of the 'deviant' individual, it may be expected that there will be even more variety of behaviour in work than there is in the simple case of group-determined behaviour.

THE SCOPE OF THE INVESTIGATION

This review of the various kinds of information required in the three parts of the series of behaviours at work can be summed up briefly by saying that what is proposed is simply a causal analysis of the last part of this series, the person's actions. The cause of action is not the situation but the person's evaluation of the situation. This evaluation is itself caused in the sense that the historical chain from action to changes in the situation, to fresh evaluations, to further action, is unbroken in either direction. It would therefore be appropriate to elaborate the account given above of the person's present situations, evaluations and actions so as to include the situations, evaluations and actions which preceded these in the person's history. This is hardly a practical possibility, both because what happened in the past can be ascertained only from the person's present report and because including even this much information in detail would make an investigation unmanageable. The most that is possible in practice is to include some report of the person's upbringing and occupational history but, beyond that, to concentrate on the analysis of present or systematic causes rather than historical causes.

However, there are more important restrictions on the scope of a manageable investigation. The systematic cause of the person's actions at work doubtless is his present evaluation of his work situations in relation to his other engagements. The account of the information that would be required to determine the person's present evaluation implies that this can be directly analysed on the basis of his report of his various covert behaviours. But it is by no means obvious that this is so. It is implied that what is to be analysed is the person's recurrent or habitual behaviour in and evaluations of a continuing situation, and not merely his momentary reactions to short-term variations in the situation. It is assumed, that is to say, that a settled evaluation of the whole set of situations at work, in relation to other engagements, is possible only if these situations at work and other engagements have been experienced as continuing and, in major respects, stable, over a substantial period; i.e. years rather than months or weeks. Even so, the assumption that either the evaluation of or behaviour in the settled recurrent situation can be sufficiently determined from the person's report, backed up by observation in the case of overt action, is at the least difficult. The evaluation that is required is complex, including numerous distinct elements. The elements, whether these are the person's position in society and aspiration for the future or the more obvious evaluations of his present work situations, are none of them the kind of process which the person is likely to evaluate, consciously and with full deliberation, unless he is asked to do so by a psychologist. It is therefore somewhat difficult to devise a clearly effective means of investigation which will so direct the person's attention to matters that he has probably never fully considered before that his present report, in the situation of the investigation, can be assumed to be really adequate. Besides this, some at least of the persons investigated may be expected to be relatively incapable of a deliberately articulated report, simply because their level of verbal (or educational) sophistication is too low. An extreme example of the combination of the two difficulties may be taken from the replies of an Australian factory process worker to certain questions about social classes. When asked: 'What do you think might be the main classes in Australia?', he replied: 'Buggered if I know.' Following this, when asked: 'To which class do you belong?', he replied: 'The booze class, I s'pose.' It is true that these responses substantially agree with other information about the person who made them, whether from his own report or from observation. It is also true, however, that if he had had more skill in the use of words (depending on more education) he would probably have been able to answer the question put in more general terms, without changing his rather narrow frame of personal reference.

The conclusion is that the investigator is likely to be committed to an elaborate reconstruction of behaviour rather than a direct analysis. In this, there is no great difficulty in principle. The information one has is in three parts: the person's situations at work, as described by the observer; followed by the person's behaviour at work and his opinions, evaluations or judgements of work situations and other engagements, as reported by him; followed, lastly, by his behaviour at work, as described by the observer. In practice, however, the causal links in the person's covert behaviour may depend on remote and inferential reconstruction rather than immediate report, as may be shown by two examples.

It is observed that some tradesmen (skilled workers, as defined by law) perform well in their jobs and are satisfied with their jobs, while others do the minimum required by their employers and report themselves as dissatisfied with their jobs. It is also observed that this dichotomy is significantly associated with the observed dichotomy among trades between clean and light (e.g. instrument fitter) and heavy and dirty (e.g. maintenance engineer, engaged on repairs and re-fitting). It is a pure inference, however, that the worker sees his situation as the observer does. The observer may conclude that some tradesmen regard themselves as well off (and as highly co-operative and highly productive) *because* they appreciate the superior physical conditions of their occupation; or, conversely, that other tradesmen regard themselves as hard done by (and are therefore relatively unco-operative) *because* they take a poor view of the necessarily unpleasant conditions of their work. But these are inferences, not observations. If the investigator has sufficient information on the intervening evaluations of the task situations, these may be taken as observed rather than inferred. In practice, it is so unlikely that the detailed evidence necessary would be available that one must assume the necessity of remote inference.

A more difficult case may be taken. It is observed that persons who are content with their place in society (however that may be defined) are significantly more likely to be content with their task and to perform satisfactorily (from their bosses' point of view) in their task. It is inferred that the second and third findings derive from the first *because* the person sees a significant relation between his place in society and his situation at work, in both cases judged by himself. Here again, this is an inference, not an observation. Unless the critical link between place in society and job satisfaction or job performance is particularly investigated, the middle term can be inserted only by inference. In practice, again, direct evidence is unlikely in the nature of the case, so that one must again rely on inferential reconstruction.

It is true that what is proposed is the ordinary comparative analysis of events, within the terms of what is sometimes called the 'natural experiment', or events as they occur, in cases in which the psychologist's deliberate intervention is impossible. It follows that, in principle, the inferences proposed are not remarkable. In any case, it is necessary for the observer to infer how the person observed perceived the situation and what were his intentions in action (cf. the author's discussion of this (1957, ch. II)). In any case, that is, it is necessary for the observer to make what MacIver (1942, ch. XI) calls a 'dynamic assessment' of the person's evaluation of the situation in which his actions have been observed. This assessment does not depend on the definition of an abstract intervening variable, though it certainly does depend on hypothetical constructs; but what it depends on ultimately is the same kind of simple analogical judgement as one necessarily makes of the behaviour of others in ordinary daily life. The fact remains that, in a complex case far removed from ordinary daily life, the judgement or inference made can only be precarious.

There remains to be considered an even more fundamental difficulty in the analysis of behaviour at work. If it is allowed that this must depend upon elaborate reconstruction rather than direct analysis, it may still be supposed that the task is simplified by concentration on recurrent or habitual or long-term behaviour. But it is a pure assumption that there is any such behaviour at work. For the single immediate event, the reconstruction of covert behaviour by analogical judgement may be simple. If one observes that the person has operated his guillotine, only one inference may be credible, that he did so because he intended to cut off the slats of metal at the specified lines. The analogical reconstruction of a coherent set of habitual or recurrent perceptions and judgements is by no means so obvious. There is no direct evidence that perceptions and judgements of the person's various situations at work (or elsewhere) are coherent or articulate, which difficulty has been sufficiently discussed. There is no direct evidence, however, that they are even habitual or recurrent; or even that the person's actions in a stable and continuing set of work situations are themselves stable and coherent. To investigate these matters directly would be an enormous though doubtless necessary task. Meantime, a practical answer must suffice: if it can be established by the reconstruction of covert behaviour that there are some plausibly coherent evaluations of the work situation, it can be assumed that these represent what persons actually do. The general need for the more direct analysis of evaluative behaviour remains.

The conclusion is that the industrial investigation described in the following chapters is necessarily incomplete because, in the present state of knowledge in psychology, it is not possible to undertake the systematic exploration of the enormous field of overt and covert behaviour set out above within the limits of one investigation. It is therefore necessary to adopt a number of practical short cuts which depend on assumptions about behaviour in general that can only be tested in the course of the investigation. This point is merely a return to the point made at the beginning of the argument, that behaviour in ordinary daily life situations has been little investigated: the consequence is that any one investigation which is in principle directed towards this behaviour is likely to be limited in its scope and hypothetical in its conclusions.

The way is now clear for the consideration of the practical difficulties of investigation, which are even more considerable. The purpose of the investigation is to study behaviour at work, even if in some critically important respects this can be studied only hypothetically and inferentially. But behaviour at work includes an enormous range of behaviours in various work situations, far more than could be included in any one investigation. The particular field or part of the range must therefore be decided upon. The field that was selected was the behaviour of persons working in factories, for two reasons. The interest of the body which provided the major financial backing for the investigation described in the following chapters was in the causes of industrial tension, or the reasons why factory managements and workers disagree. This coincided with the interest of the investigators in studying the behaviour of factory workers because, if empirical psychology has been little concerned with ordinary life behaviour in general, it has been even less concerned with the behaviour of manual workers, verbally unsophisticated persons, or relatively uneducated persons. In short, psychology has been little concerned with the behaviour of anyone other than sophisticated middle-class persons, whether juvenile or adult. An obvious gap to be filled, therefore, was the undertaking of the deliberate study of factory workers (together with their bosses, as far as possible); but a generally directed study of behaviour seemed appropriate, and not merely another study of productivity, or adequacy from the point of view of the factory management.

The definition of purpose, however, raised a number of difficulties. Those that were due to the particular local conditions under which the investigation was done are considered in chapter II. The more general, which were inherent in the nature of the proposed investigation, must be reviewed here.

For the purpose stated, the appropriate general method of investigation was the biographical interview, backed up by observation where necessary. These are the methods that are most likely to give a wide range of information in a relatively short time (cf. the author's discussion of this (1957, ch. IX)). The interview in particular is specially suitable for gathering the information necessary for the analysis of the person's reconstructed evaluations of his situations or engagements; and it seemed reasonable to concentrate on this, the intervening causal process or middle term of the series of behaviours, because this is the least studied and therefore the least known term. The combined procedures adopted were therefore time-consuming but straightforward in principle.

In the review of the person's situations or engagements, some difficulties were inevitable. For his task, his own report of what he is supposed to do is almost sufficient and needs backing up by observation only to the least extent which is necessary for the investigator to be able to attach a concrete meaning, derived from observed operations, to the worker's description of these operations. The pay system, being a public and official fact, is sufficiently defined by the statements of the factory administrators who are concerned with it. The general social organization of the factory is less easily defined, as it is likely to be an inference made by the investigator rather than an obvious public fact. Even so, the inference should be easily and unambiguously made from the statements of members of higher management and of foremen about how the factory is run. With the behaviour of the person's bosses, the first difficulty appears. The simple counterpart to the person's statement of how his various bosses behave is their own statement of how they behave. But, for various reasons (which are detailed in chapter II), it was impossible in the case taken and probably would be impossible in any case either to interview all the persons concerned, whether foremen or of higher supervisory grades, or to get reasonably complete information about actual supervisory practices even from all the bosses who were interviewed. In this case, then, the analysis of the person's situation depends on incomplete information. A similar but more marked difficulty occurs with the definition of the person's relations with his workmates. From his point of view, they are as he states them in the interview. The least counterpart of this information which would be necessary is his workmates' statements of their relations with him. But this would require that all his workmates should be interviewed. As this was impossible in practice (see chapter II for details), there is no systematic evidence from, as it were, the other side; the only evidence available is in a few cases where it happened that most or all of the person's workmates were interviewed.

It is true, of course, that what could not be done by interview (for local and practical reasons) could, in principle, be done by observation. The extended and systematic observation of the behaviour of various bosses at work and of the behaviour of actual working groups or gangs in the factory should provide enough evidence of the situation as seen by the observer. However, this is not a practical possibility. In any industrial situation, it is not very likely that the presence of a detached observer whose sole function is to watch men at work would be tolerated for anything like a sufficient period for the definition of these situations from observed behaviours. Besides this, the time required would be prohibitive in any ordinary case that was limited by finance. Both difficulties apply even more strongly to the observation of the person's engagements outside the work situation. The sociological facts of his occupational history and family situation may be taken as stated by him, though the former might, in principle, require independent confirmation. It has already been argued that there is no practical possibility, in any case at all, of actually observing his class, family or unprescribed behaviour. The point is therefore repeated that these can be treated, at present, only as perceived behaviour as reported by the person.

In the review of the person's actions at work, the difficulties inherent in the nature of the case become insuperable. In this vast field of ignorance, the most one can do is to reduce to exemplary detail the general account given above of the matters that would require investigation. In the performance of his task, the person may do it exactly as it is laid down officially or he may introduce individual variations or short cuts; and he may do it to the maximum of which he is capable or to the minimum which is acceptable. His relations with his immediate boss may be generally co-operative, or argumentative or obstructive. In the latter two cases, one should enquire into the occasions which give rise to argument or obstruction, and into the manner in which either of these is conducted, and into the effectiveness of either. Lastly, the person's relations with his mates may be cordial or variable or generally hostile. The important questions here would be: with whom does he crack jokes or spend his tea-break or lunch hour, then at whom does he swear occasionally, and on what occasions, and then with whom does he deliberately not co-operate, on what occasions.

In principle, it would be possible to get information over this vast and relatively unexplored range; in practice, it would be difficult in any case, regardless of local field conditions of work. The three possible sources of information may be considered in turn.

(i) The person's own report should include some statement of his actual behaviour at work. Doubtless: but the report made can only be made in very general terms. There are some important details which hardly could be investigated in a general interview: e.g. on what occasions has he actually quarrelled with his boss or with his workmates, or what does he actually do in his task which is against the officially laid down task procedure. Further, even if these and other less ego-involving activities could be the matter of direct questioning, the interview procedure would become unmanageably long and detailed. It may be concluded that the investigation of reported behaviour at work is a more than substantial topic on its own.

(ii) A necessary counterpart of the person's behaviour at work would be the investigator's observation of this behaviour; for it cannot be assumed that there is a simple correspondence between what the person does as he perceives and reports it and what he does as a detached observer perceives and reports it. It has already been noted that a bare minimum of observation was possible and, one can say, would be possible in any case. But it must be allowed that this minimum falls far short of what would be necessary for the systematic checking of the person's reports of his behaviour. This would require that he is observed at work to a sufficient extent for the investigator to be able to say, for example, in just what respects he customarily does the job his own way rather than the official way; or over what sort of issues he is at cross purposes with his immediate boss or his workmates. In any ordinary case, it is not likely that the person would be agreeable to having an observer so closely trailing his actions, both on and off duty. Besides this, the time required for observation on this scale would be vastly greater than the time required even for extensive interviewing. The conclusion is, again, that the observation of behaviour at work is a more than substantial topic on its own.

(iii) The reports made by bosses of various grades on the behaviour of the workers would provide some counterpart information. It may be that this information would have to be checked by observation, for the actions of the worker as perceived and reported by the boss are not necessarily the same as those actions reported by a detached observer. Beyond this, however, there are two more obvious difficulties. The reports that would be necessary would require the detailed discussion with each boss, of whatever grade, of the actual behaviour of a number of specified individuals. It is more than unlikely that information of this kind could be obtained from a good interview unless the interview time were unlimited and the patience of the person interviewed inexhaustible. Besides this, it is unlikely that more than a very few bosses would be able to make firm and definite statements about the details of behaviour of their workers. The boss's function, at any level, is to get out production; and the consideration of individual differences and idiosyncrasies among producers is secondary to this. The consequence is that many bosses, especially foremen or immediate supervisors generally, whatever their title, are likely to be incapable of a detailed evaluation of the performance of their workers. Many do not have the verbal sophistication required for such an evaluation and for almost all it is not necessary in the course of their duties. Bosses (supervisors) of various grades were interviewed, and their general opinions of their workers and their problems and general methods were recorded and analysed. But the general conclusion is the same as before: that the way the boss sees the worker in day-to-day activity is a substantial research topic on its own.

There remains the possibility that evidence of overall adequacy of behaviour at work (in the simplest case, of productivity) could be used. This is the conventional evidence of applied psychology, even though it is not very often put forward; but here too there are considerable practical difficulties. Even if it were available, this evidence would not be particularly relevant to the purpose of a general investigation of industrial behaviour. Evidence of the worker's satisfactoriness from the point of view of the management is necessarily inconclusive. It is quite possible, for example, for a worker to be satisfactory from this point of view (punctual, exact and hard working) and to be unsatisfactory from the point of view of his workmates, or the other way round. Besides this, the evidence of the worker's general adequacy may be doubtfully reliable or complete. For some tasks, there may be production records, but there are not so very many tasks for which individual production or the production of small groups can be measured. In any case, productivity is not a satisfactory measure of general adequacy because it often depends on local or temporary factors of supply, layout and equipment that have nothing at all to do with the individual worker's effort. Much the same applies to the estimation of what one might call residual or background adequacy, from factory records of absences from work, stoppages on the job and accidents. Even if these records are meticulously kept (which is not usual in Australian factories) they cannot give more than a very approximate estimate of the worker's adequacy because in this case too there are many important factors that are altogether outside the worker's control. The last and most obvious kind of evidence that might be used is the foreman's ratings of adequacy. These have so often been shown to be unreliable that the point hardly needs supporting with particular evidence. It may be that the conditions under which estimates of this kind would be reliable could be simply defined. But this, again, would be a major research project on its own; and, until it is done, it cannot be assumed that this sort of evidence of behaviour at work is any more usable than are the more abstract kinds of evidence of general adequacy.

CONCLUSION

The consequence, then, is that the actual scope of the investigation is much less than that of the general plan set out above. The person's evaluation of his work situations and other engagements will be studied but his actions and their relation to his evaluations will not. The restriction, if it is unfortunate, is inevitable because so little is known about the second and third parts of the series of work behaviours that studying one by itself requires a large-scale project. Given this, it seems reasonable to start with the second part.

It is true that there is thus no direct evidence that the main content of the investigation is what it was supposed to be. What the person acts on is his final practical judgements of the whole set of his engagements. It is suggested that if he cannot directly state these judgements or analyse their derivation, the comparative analysis of the judgements which the person can report, in relation to various situations, should enable the investigator to make a coherent reconstruction of the whole course of the person's covert behaviour. But the investigator can assert that it is this behaviour, as reconstructed, which leads on to action only if he has actually shown a set of relations in action. It is equally possible that what he has reconstructed is no more than a set of momentary, forced and unconsidered opinions. For most persons, and especially for factory workers, sitting down with a psychologist to discuss one's work is an unusual sort of activity. In the course of their ordinary activities, again, most persons probably do not often stop to consider their work in general, or their place in society, aspirations for the future and so on. It is therefore quite possible that the report made in the interview is honestly made and is true at the moment when it is made, but bears little relation to the person's long term settled and practically effective views. Because these are quite likely to be inarticulate and incoherent, a brief interview may be a quite inadequate way of elucidating them: and the only satisfactory check is to test them against the person's actions.

While this is strictly correct, the practical difficulty must not be exaggerated. The extreme possibility that the person's opinions, as he states them in the interview, are little related to his settled and effective views seems unlikely in the light of common experience and (in the actual case) unlikely in view of the coherence and credibility of the finally reconstructed set of evaluations. In any case, it can be written off as a general possibility if it can be shown, from whatever evidence, that there is some relation between opinions and actions. There is of course no question of stating a whole range of specific relations, as too little is known in this field, but any observed cases are relevant. Further, it is not necessary that the relation should be close. In any situation in which there are several possibilities of action, behaviour is likely to depend partly on experiment and may therefore not derive uniquely from the evaluation of the situation, which is to some extent inconclusive.

The observed relation between job satisfaction and productivity may be considered first. Katz and Kahn (1952, p. 662) report an association between responses to the question: 'Do you feel you are really a part of your work group?', and group productivity. The men who felt that they were a part of their group tended to be the higher producers, the association is highly significant for the large number of cases taken (2,085 factory workers) and the coefficient of association, ϕ, is 0·22. In the same place, there was an association between the foreman's opinion of the internal co-operativeness of his group and its productivity: $\phi = 0{\cdot}13$, so that for 316 foremen $P < 0{\cdot}05 > 0{\cdot}02$. The observed associations, then, are quite low; but the more remarkable fact is that they are significant at all, for three reasons. It seems unlikely that the measure of productivity was accurate, or was even on a uniform basis, for the whole of the large sample. Besides this, productivity obviously depends only in part on job satisfaction (of whatever kind) and also in part on physical factors over which the worker has no control. Lastly, the single questions asked, if they are accurate measures of an aspect of job satisfaction, are very much less than a coherently reconstructed evaluation of this aspect. Heron (1954) reports a product-moment correlation of 0·353 between job satisfaction and satisfactory work, for 144 bus conductors. The point of this case, with a higher satisfaction for a much smaller population, is that both variables were carefully and elaborately constructed from a number of elements. There would be little point in going to review the scattered literature on the relation between job satisfaction and production. As Viteles notes (1954, p. 11), findings in this field are inconclusive; but the point at issue is not the nature or degree of the general relation. The point is simply that workers' opinions have been shown, if only in these two cases, to bear an intelligible relation to their behaviour. The general possibility that their opinions are merely artefacts of the interview may therefore be discounted.

The same point may be made by considering some of the interview material itself, the reasons given by 127 factory workers for leaving previous jobs. This material is discussed in chapter III: here a statement of its general nature is sufficient. There were four main kinds of reasons: none in particular, just wanted a change; the job came to an end itself (the contract was finished or the firm bankrupt); family reasons (moved to be near parents, relations); and job reasons, or dissatisfaction with the task, pay, bosses, or workmates. Now the last kind of reason was given far more frequently than the first and rarest kind: so obviously so that no statistical test of significance is necessary. It may be concluded that if the men interviewed merely invented their reasons retrospectively, they worked surprisingly hard to do so in a situation on which this was not obviously necessary. Further, when dissatisfaction was mentioned it was usually backed up with specific detail: the task was dull, dirty, hard or dangerous in fully specified respects; the pay was not up to the specified responsibilities of the job; and so on. Of course one has only the worker's word for it that he did make these judgements and then act on them. There is no actual evidence that he did not leave his job, on whatever grounds, and think up the reasons given only in the interview. In the nature of the material, however, this possibility is incredible. Here again, then, is a case in which practical judgement of the work situation is related to action. One would not expect, of course, that the relation between dissatisfaction and action would be close, if only because of the many other factors that are relevant. But it is clear that the relation would be effective and here again the most important point is that the practical judgement on which action may be taken is very much simpler than the worker's elaborately reconstructed evaluation of the situation.

The fact remains, of course, that the second part of the investigation has yet to be done. If it can be assumed that intelligible relations between evaluation and action would be found, nothing at all can be said about what these relations would be. The main value of the investigation is in showing how the reconstruction of covert evaluative behaviour can be undertaken. If, even in this respect, the results that are discussed in the following chapters are more indicative of further hypotheses and further fields to be explored than they are of firm and generally relevant findings, that is no more than the inevitable deficiency of work in substantially new territory.

CHAPTER II

THE FIELD OF THE INVESTIGATION

THE general principles or general point of view of the investigation have been discussed and notice has been taken of the major limitations of scope that had to be accepted in practice. The next step is to review in detail the conduct of the investigation, or what was done, with whom and how.

In many cases it is not necessary to detail these matters. It may be obvious that, with regard to the criteria taken as relevant, the persons studied were representative of the general population. The purpose of the investigation may have been evident to the persons who took part in it if they were highly sophisticated volunteers: e.g. students of psychology. Alternatively, even if the purpose was not evident, the tasks set may have been accepted without comment as being in line with the person's usual activities as, for example, with investigations requiring a school class to do something like an ordinary class exercise. In either case, there would be no need to detail the methods used to convince the persons taking part that their doing so was worthwhile and relevant to their purposes because no specially elaborate methods would be required. Then again, if the methods of investigation used were the conventional methods of the psycho-physical laboratory, no detail is required beyond a reference to a laboratory manual. Here too, little detail is required of the instructions given to the persons taking part, for it can be assumed that the task requires no special intellectual or verbal sophistication. Lastly, if the results of the investigation were analysed by the ordinary methods of parametric statistics, no more than a statement of the actual statistics used is necessary.

But this is, in effect, a summary of the conditions of work of the experimental laboratory. An investigation in this field could not be more different from one which is directed to ordinary daily life behaviour for the most part and to the behaviour, at that, of persons who are for the most part intellectually unsophisticated. As the investigation was planned, it became clear that several practical questions required careful consideration. The chief of these were: how to persuade factory managements and workers that it was relevant and worthwhile and, linked with this, whether it would be possible to study a fairly representative sample; how to arrange and conduct the interview and how much observation to do; and lastly, what methods of analysis to use. The most convenient way of showing how these questions were handled is by a review, in historical order, of the introductory and detailed planning stages of the investigation. The first consideration is how the project was represented to various factories, and on what terms was it accepted. The next is a general description of the persons investigated. The last is a description of the methods of investigation and analysis. The full account is necessary in order to make it clear that the obvious difficulties of the 'natural' experiment were by no means overlooked, either in the field procedure or in the drawing of conclusions, as set out in the following chapters. The reader who is prepared to take these difficulties for granted, however, may prefer to go straight on to the last section of this chapter, the *Summary*.

THE FACTORIES THAT WERE STUDIED

From the simple point of view of laboratory psychology, the first move in an investigation of behaviour at work is to collect a sample of persons who work. What is proposed is a 'natural' rather than a 'true' experiment: that is, a comparative analysis of variations in behaviour concomitant with variations in conditions of work, of the sort which is usually found in sociology, and not a laboratory investigation as usually conducted, regardless of the person's context, by the experimental psychologist. Even so, it would be appropriate to define the composition of a general sample of persons who work and then to locate a number of appropriate persons and study them. In practice, it is not likely that one could do this. Most of the studies of industrial behaviour that have been made have been undertaken from the point of view of the management: that is, have been directed to the particular practical consideration of ways in which management could increase production. It is conceivable that an investigation could be undertaken from the point of view of the workers, though this has apparently not happened. On the face of it, then, the easiest way to get access to a number of workers would have been to propose the solution of practical problems, as required. But a general industrial population could not have been got in this way. In Australia, there are still many factory managements that are unaware even that management requires study and analysis and not only experience. The number of firms that are aware of their problems to the extent of having even a vague interest in industrial psychology is very small. These firms are certainly not representative of Australian factories in their social organization, and probably not representative even of the major industrial groups or types.

This line of approach was therefore discarded (except in one case): without regret, as the adoption of a general line made it possible to meet a critically important requirement, that of independence. The initial approach to factory managements and the later approach to the workers stressed the same four general points. Firstly, very little is known in detail of actual daily labour-management relations and even less is known from the workers' point of view, so that many attempted solutions of practical problems in the factory depend only on ignorance illuminated by intelligent guesswork. The university is the only body which is capable, in principle, of exploring this vast field of ignorance. Thirdly, however, if it does so it can make no promise that the results of the enquiry will be relevant to any specific practical problems. The last and most important point follows: that the investigation was proposed by a university department, as an independent body, for its proper purpose of advancing knowledge. The investigators, that is to say, asked the organizations and individuals approached to help them, but held out no promise that any help could be given by them with particular problems. This point was emphasized by making the subsidiary point that the only report of the investigation would be a published report available to everyone.

The introduction of the investigation was in seven stages. In the first, the general idea was presented to a number of firms which, on general acquaintance with local industry, seemed likely to be interested. Most of these declined at once or made no reply. With the remainder, the investigation went on to the second, third and fourth stages. These were the discussion with management of the detailed aims of the project; of the methods of work proposed and the time required; and of the introductory procedure which would be necessary. At the end of these stages, several firms declined co-operation. The fifth to seventh stages were finally gone through in the seven factories that did co-operate. These were the explanation of the project, in general and in detail, to foremen, trade union officials, and lastly the workers themselves. In all these stages three practical points were specially emphasized. Firstly, that the investigators were working by permission of the management but in no sense for it. Secondly, that the project could go on only with the permission of the foremen and union officials but, again, was not undertaken for them. Lastly, that anyone who volunteered for interview would be interviewed on the firm's premises and in the firm's time. When necessary, a subsidiary point was made with foremen and trade union officials, that it must be left to the investigators to get volunteers. Offers of assistance in this were appreciated but declined, in the interest of maintaining the strictest and most obvious independence.

This was the procedure in principle. Elaborate as it was, no simple procedure would have sufficed to gain the interest and active co-operation of the persons concerned, whether management or workers, because the investigation proposed was something entirely out of their experience or even their imagination. However, it was not possible to follow this procedure exactly in all cases. Certain consequences of the practical difficulties encountered in various stages of the introductory procedure remain to be considered, in relation to the population that was studied, as shown in Table 1. There are five important points:

TABLE 1

The population interviewed

Factory	Product or type of industry	Fully interviewed				Interviewed as consultants	
		Men	Women	Foremen	Professional employees	Foremen	Higher management
A	Chemicals	58	—	9	14	—	2
B	Chemicals	10	—	1	1	—	—
C	Sheet metal goods	35	5	—	—	4	4
D	Sheet metal goods	18	—	4	—	—	2
E	Repetition engineers: various products	6	5	—	—	3	3
F	Clothing	8	43	—	—	3	1
G	Processed food	62	32	—	—	6	7
	Total	197	85	14	15	16	19

(i) It is clear that this collection of seven factories is in no way representative of industry in Australia, in the state of Victoria or in the Melbourne urban area. Large industrial groups, such as mining, building and wharf labouring are not included at all, and engineering (production, not maintenance) is hardly included. The point need not be laboured beyond this, though it may be observed that the specification of the products of the factories is deliberately vague, in order to reduce the possibility of identification in the local community. A more exact specification would show even more clearly that the group studied is not representative.

(ii) This applies particularly strongly to factory F. Its relation to the company which controls it is unique, in respects which cannot be specified for the reason given; so that it is quite untypical of Melbourne clothing factories.

(iii) In factory G alone, the investigation was carried out at the request of the management, in order to discover possible reasons for labour turnover, and not as a general research project. Besides certain consequent differences in the introductory procedure, there were necessarily differences in the range of the interview, which was much more concentrated on affairs inside the factory than it was in other cases.

(iv) In any case, the range of the interview depended on the grade of the person being interviewed. Factory workers, both men and women, were all volunteers with whom a full biographical interview was therefore appropriate. With other persons, it seemed in some cases undesirable to invite their co-operation to this extent, or, in some cases, obviously unnecessary. It seemed clear that many foremen, for example, would have been reluctant to refuse a definite invitation to take a full part in a project that had the support (as they often, if somewhat wrongly, saw it) of their higher bosses. The conservative rule was therefore adopted that all persons of the grade of foreman or above should be interviewed in the first place as expert consultants on the management methods and problems of their firm. This, of course, was necessary as a counterpart to the workers' information on these topics. Beyond this, however, a full biographical interview would be attempted only with foremen and other immediate supervisors (professional chemists, in factories A and B, were the only others in question) who were clearly willing to go this far. The last two columns of Table 1 therefore include a number of foremen with whom the purely consultant interview was found appropriate, as well as members of higher management, with whom a fuller interview would have been irrelevant.

(v) The last and most important consideration is that the introductory procedure, as it was actually carried out, varied in accordance with the management's perception of its relevance, in relation to its own immediate problems of production. Factories A and B were exceptionally well placed to give the time required for an elaborate introduction and in both the management (common at the higher levels, as B is a branch of A) was appreciative of the general aims of the project (partly in consequence of certain production difficulties) and also of the need for a detailed introduction. In both these factories, therefore, the whole introductory procedure was carried out as laid down, under good physical conditions and under no pressure of time. The same was generally true for factories D and F, except that their physical layout and conditions of production made it inevitable that the last and most critical stage of the introductory procedure, explanation of the project to the workers, had to be put over under some pressure of time (in a few minutes before the dinner break) and under some pressure of space (on a crowded factory floor). The same sort of difficulty was found in factory E. Here, the management was entirely co-operative but the last stage of the introduction had to be abandoned because of the physical impossibility of talking to more than a very few of the workers in any one place. In this case, therefore, the last stage was the explanation of the project to the charge hands of the factory (who were technically members of the staff, but who themselves volunteered to be interviewed) and the forewomen (who made themselves responsible for getting women volunteers). In factory C, a different kind of difficulty was found. The project was accepted by the management at the instance of the personnel officer, who had not been in the factory long, was little known, and had insufficient contact with other members of management. In consequence, there was no difficulty with the earlier stages of the introductory procedure, but the last stage could not be satisfactorily handled. The most that was possible was to make a rather formal explanation of the project, over a public address system, in the workers' lunch hour, in the factory canteen which only about a third of them used. The procedure in factory G was the least satisfactory of all, as the last stage had again to be omitted altogether. This factory has an exceptionally large number of small sections, so that explaining the project to the workers concerned would have been a major exercise on its own. The introductory procedure therefore went through only the first six stages, to the explanation of the project to a routine monthly meeting of the firm's foremen (the union officials had already been consulted). But what was explained at this meeting was not a general research project: it was a proposal to investigate the particular local reasons for labour turnover, including by implication the foreman's efficiency in retaining his workers. It had to be left to the foremen to get volunteers for interviewing, and the obvious consequence (as detailed in chapters VI and X) was that these were more often the foreman's nominees than they were really volunteers.

The point has already been made that the factories studied cannot be regarded as a sample of industry. The second point may now be made, that the persons interviewed were not in all cases representative of the factories in which they worked, with regard to the sociological criteria that could be determined from factory records. For the men interviewed, factories B and E can be disregarded, as both are too small for sound statistical comparisons to be possible in any case. Factory F must be disregarded, both because of the peculiar features noted above and because most of its workers were women. For factory G, in which only the number of men employed in various sections could be reliably determined from the records, it is none the less clear that the men interviewed were no sample of the factory's workers. About 10 per cent volunteered to be interviewed, which was the lowest rate by far in any factory; and the whole of this fraction came from less than a half of the manufacturing and maintenance departments, so that more than a half of the factory was entirely unrepresented. For the remaining three factories, the records were sufficient for account to be taken of age, length of service, section in which employed, general grade of employment (skilled, semi-skilled or unskilled) and, less reliably, country of origin (Australian, British, foreign). In factories A and D, in both of which 24 per cent of the workers were interviewed, this fraction was quite closely representative of the whole body of the workers in all of the respects mentioned. In both, there was a slight and barely significent deficiency in one group of workers only: the outposted labourers, or the men who work in relative isolation on general cleaning-up jobs (yardmen, etc.) in any part of the firm's premises, as required. In factory C, where 18 per cent of the workers were interviewed, there was only a fair representation of the whole body of workers. The proportion of skilled men (tradesmen) to labourers was correct, as were the proportions of particular sections within these main groups. There was a highly significant deficiency, however, in all groups of semi-skilled workers, or machine operators.

With the women workers, only four of the seven factories need be considered, as factories A, B and D employed no women, or at the most, a very few who worked in the canteen. Of the others, factories C and E must be disregarded, again because sound statistical comparisons cannot be made with the small numbers interviewed. In factory F, 29 per cent of the women workers were interviewed and this group is closely representative of the whole population except only in occupational sections: there is a highly significant deficiency of tailoring machinists. In factory G, only 5 per cent of the women workers were interviewed and these were quite unrepresentative of the occupational sections of the whole factory, in the same way as for the men, and for the same general reasons set out above.

For the foremen and men of higher supervisory grades, the question of representativeness cannot be raised, as no attempt was made to include a wide range of persons who were interviewed in detail. There remains one important practical point. In some cases, as noted above, the persons interviewed were closely similar to those who were not interviewed in the length of their service in the factory. This is true for a rather coarse break-up of length of service but it is not true for the extreme division between very short service (two months maximum) and all longer periods. In no factory was there a proper representation of persons with very short service. It has been sufficiently noted that most labour turnover occurs very early in the person's engagement in the factory, and drops rapidly in weeks and months to a steady and very low rate. An exact specification has been proposed Rice, Hill and Trist (1950), who separate the labour turnover of a relatively progressive and well organized firm into two parts: the men who leave before the 'induction crisis' (i.e. within their first three weeks in the factory), and the relatively few who leave after this. The point is of practical importance. No persons were interviewed whose 'induction crisis' was still before them; and, as all the factories had fairly high rates of turnover, it must be assumed that these persons simply did not volunteer for interview. In the terms of the investigation this is not surprising, but it follows that the conclusions drawn must be limited to the relatively settled workers.

With regard to the factories, the social organisations of which are discussed in chapter X, a conclusion can now be stated. For the general features of social organization, the reports of members of higher management and of foremen (however they were interviewed) are probably sufficient. The adequacy of the counterpart information from the workers, however, depends on the case. For the men, comparisons between factories A and D may be relied on: factory C can be included with reservations; and factory G can possibly be included if its peculiar features are capable of interpretation. For the women, the only possible comparison is between factories F and G: and the latter can be included, again, only in so far as the information available is interpretable in the light of information from other sources in the factory.

THE PERSONS WHO WERE INTERVIEWED

It is clear that conclusions about the factories studied, as distinct social organizations, can be drawn only with considerable reservations. The possibility remains, however, that the persons interviewed were fairly representative of a general industrial population in Australia, allowing only that none but relatively settled workers were interviewed, but taking the persons interviewed as one group, regardless of where they were working. This possibility obviously depends on the procedure for introducing the investigation. The very last step in this, at the very end of the seventh stage, was to point to the ballot box which was hung up in some conspicuous central position (and which was conspicuously labelled: 'University of Melbourne') and to invite the persons who were interested to write their names on the forms provided and put them in the box. The investigator's next step was to collect these forms over the weeks that he was engaged in interviewing in any one place and to arrange interviewing time-tables. The procedure worked much in the same way in all the five factories in which it could be used (not E or G); but two slight differences may be noted. In factory C only, there was a very high rate of return of persons who were nominated for interview by their workmates, a fact that was clearly associated with the general conditions of introduction into this factory. Most of these persons were quite willing to be interviewed none the less, and in this factory alone the flow of volunteers was partly kept up by asking workers who had been interviewed (but not foremen or union officials) to urge on their mates. Besides this, it was observed that the rate of volunteering among women in all factories (including even factory F) was significantly lower than the rate among men. As a cultural phenomenon, this is not remarkable. Together with the observed fact that the final basis of volunteering differed between factories, it does raise the practical question: what were the general reasons for coming forward for interview?

The answer can be only roughly inferred from insufficient data. In factories A to E, there was provision in the interview check list for asking, in general terms: 'What did you think of the interview?' The question was not put in all cases. It was not put where it was obvious that the person had come to be interviewed because he wanted advice on his personal problems or where it was obvious that he had come only because he did not like to refuse. In consequence, specific statements were made by a majority of the persons interviewed in these five factories but the outlook of the remainder can only be inferred from the interview record as a whole. This last point applies to all the persons interviewed in factories F and G, in both of which the direct question was never asked. A very approximate summary is therefore the most that can be given, for the men and women interviewed (N = 282), as shown in Table 2. Two points may be noted about this table.

TABLE 2

Stated and inferred reasons for consenting to be interviewed

Type of reason	Percentage
To increase knowledge, say something useful, help the investigator	58
To do the right thing by the boss or the firm, represent their point of view	17
To have a rest *or* to please the investigator (persons who were not really volunteers)	13
To get advice on various personal problems	12

Firstly, the categories are not exclusive, so that the person's explicit statement might include points under two or even three heads. The more important point is that the conservative rule was used in tabulation that any statement which could possibly be put in one of the last three categories (whether by omitting a conjoint reference to the first or by simple inference from the whole record) was so placed. In consequence, the table presents a most conservative estimate of the percentage of workers who did appreciate the main point of the investigation and a deliberate exaggeration of the percentage who took part in it for relatively secondary reasons. It may be concluded that, in spite of the local differences in the introductory stages that were reviewed above, the investigation was quite efficiently put over and quite well accepted as something that mattered in its own right.

A general statement of the sociological characteristics of the persons interviewed may be made next. The distribution of ages is shown in Table 3, for all persons who were fully interviewed.

TABLE 3

Ages of the persons interviewed

Age range	Number of persons		
	Men	Women	Foremen and professional employees
Less than 20	8	21	—
20–24	28	12	8
25–29	31	8	4
30–34	32	7	3
35–39	27	10	5
40–44	20	9	2
45–49	18	7	1
50–54	13	6	2
55–59	8	3	2
60–64	10	2	2
65 and over	2	—	—
Total	197	85	29

From this table, it is clear that there are differences between the three groups of persons, the consideration of which, however, must be deferred to chapter X. The point can be made here that both the men and the women range widely enough in their age for it to be conceivable that they are something like a general industrial sample. The same point can be made for length of service, as shown in Table 4. Here too, apart from the differences between groups, it would seem that there is reasonable variety among both the men and the women, except only for the point already noted that no persons with very short service were interviewed. Nationality, in the sense of country of origin (i.e. the country in which

TABLE 4

Lengths of service of the persons interviewed

Length of service	Number of persons		
	Men	Women	Foremen and professional workers
Less than 6 months	31	12	—
6 months to 2 years	67	39	7
Over 2 years, up to 5 years	39	14	6
Over 5 years, up to 10 years	27	9	8
Over 10 years	33	11	8
Total	197	85	29

the person was born and in which he lived until at least the age of five) is shown in Table 5. In this case it is particularly difficult to say how far the persons interviewed might be equated with the

TABLE 5

Country of origin of the persons interviewed

Country of origin	Number of persons		
	Men	Women	Foremen and professional workers
Australia	154	70	23
Great Britain	31	7	4
Continental countries	12	8	2
Total	197	85	29

industrial population of Australia. But in this case too it can be said that there is sufficient variety for some conclusions to be possible about differences that may be associated with national origin. The last consideration is the person's occupational grade or sub-class, as set out in Table 6, for factory workers only. This table requires some definitions, as the briefly stated categories are referred to throughout the following chapters. For the men, the classification is almost entirely as it is defined by industrial law, interpreted psychologically where necessary. Thus a tradesman is a man who has served an apprenticeship and gained legally recognized qualifications as an engineer (or mechanic), carpenter, electrician, plumber, etc. But this category also includes

TABLE 6

Occupational grade of the persons interviewed

Occupational grade	Number of persons	
	Men	Women
Tradesmen	71	3
Process workers	83	64
Labourers	43	18
Total	197	85

war-time dilutees (mostly engineers: i.e. mechanics) who do not have the qualifications required by law but whose experience was sufficient for their acceptance as tradesmen. It also includes a very few men who were working as tradesmen (e.g. charge hands whose main task was setting up turret lathes or making tools for turret lathes) but whose qualifications would not be recognized officially. The criterion, that is to say, seldom differed from the official definition; but, when it did differ, this was because the man was classified on account of what he was actually doing and not on account of what he was officially qualified to do. The same holds in general for the other two categories. A process worker is a man who has no trade qualifications but who has learnt a complex set of particular machine operations such that he can be replaced only at the cost of training his successor over some weeks or even months. A labourer is a man whose task requires only the elementary and almost universal physical capabilities and who is therefore replaceable at very short notice. With both these categories, the official classification was followed for the most part, but was disregarded in a few cases in which the evidence showed that the task had been overrated (it was only a labourer's task actually) or had been under-rated (it required considerable and specific learning on the job, although it was classified as a labourer's task). For the women, the same general procedure was followed but was more difficult to apply in practice because women's occupations in industry have a very restricted range. Two conservative rules were therefore applied. Firstly, if the task was one that only nominally requires an apprenticeship (as in the clothing trade generally) it was not regarded as a trade. Only if it was one that is done equally by men who are officially accepted as tradesmen was it so classified. Hence the very small number of women (all of them qualified tailors) who are entered as having trade qualifications. Secondly, no task performed by women was regarded as entirely unskilled unless it was quite clear from the records that anyone could learn it in a very short time. High speed repetitive work is common among women's industrial tasks; but many tasks of this kind required sustained concentration (if only to keep up with the machine) well beyond the extent of which all persons are capable. Here too, therefore, the official classification was often disregarded in favour of the psychological estimation of the requirements of the task: hence the relatively low number of women who are tabulated as labourers. The general point follows: here again, it can hardly be said that the distribution of occupational grades resembles that of the Australian industrial population. For the men, on the contrary, it seems likely that too many tradesmen and not enough labourers were interviewed. For the women, a direct comparison is hardly possible. For both, however, it can be said that there is sufficient variety among the persons interviewed for some conclusions to be possible about the differences that might be associated with occupational grade.

The conclusion, then, is that the persons interviewed cannot be regarded as a general industrial population. How far this may limit the value, or general applicability, of the findings reviewed in the following chapters is a question that must be deferred to the end of the review. However, the general point is worth repeating that there is considerable variety among the persons interviewed. It may be hoped, therefore, that comparative analyses by age, length of service, nationality and occupational class and, as far as possible, by sex and grade, will give some indications about the generality or specificity of the findings. But one more reservation must be made: that there is no way of determining how far the persons interviewed were selected on what might be called personal or psychological grounds. The point that matters here is that they were not a sample selected by the investigators, whether efficiently or inefficiently; they selected themselves, or, in two factories, were selected by their foremen. The latter process happened almost entirely in factory G, so that comparisons between factories should suffice to show in what ways it might affect the findings. In the other factories a fraction of the workers, at the most 24 per cent, came forward for interview. In some cases these persons are comparable with those who did not volunteer with regard to simple sociological characteristics; but these, the only ones for which comparison was possible, may not be the important characteristics. It is possible that the volunteers included an excess of persons who had problems they wished to discuss. The estimate that this was true of 12 per cent at the very most hardly suggests that there was an excess, but no test is possible. Again, there may have been an excess among the volunteers who were strongly for or against the management, of strongly conservative or radical political views; or an excess of persons of middle-class origin, or of working men trying to rise in the social scale. The evidence on politics and class that is presented in chapter VII shows only slight differences from that reported in other sources; but this does not directly rule out the possibility that there were differences inside the factories. Lastly, the fact that some volunteered and some did not is itself the main difference. In a sense, the investigators were very successful in getting volunteers. The main ground for taking part was the same as the purpose of the investigation, simply in order to advance knowledge. This is by no means an obvious idea, especially in a country in which psychology is little known, so that its being accepted at all shows that the introductory procedure was effective. The fact remains that, in the best case, for every worker who volunteered to be interviewed there were three who did not. Whether this difference would be associated with definable personal qualities, such as intelligence, sophistication, curiosity or any other is, of course, impossible to determine.

The last point that must be mentioned about the persons interviewed is that the interview procedure itself varied from time to time. In factory G, as the investigation (carried out in 1950) was directed to the causes of labour turnover, the information gathered was almost entirely about behaviour within the factory. In consequence, these records are of limited usefulness. In factory F, which was investigated in 1951, the interview procedure was the same in principle as the final form (for factories A to E, in 1952) which is discussed below; but there were some important differences and there was the general difference that the interview was laid out more like a rigid and detailed questionnaire than an open procedure. In consequence these records also are of limited usefulness. The number of records that were used in the analysis set out in the following chapters therefore depended on the purpose of the analysis, and in most cases it was less than the total number specified in Table 1 or Tables 3 to 6. It must be noted, however, that the analysis of effectively complete records only (the numbers and sources of which are clearly specified in the following chapters) is not likely to vitiate the main conclusions for the male workers.

These conclusions depend on 127 records only (i.e. factories A to E only) for the most part, on 189 records for some points and on all 197 records (including factory F: the reasons for disregarding it generally are stated in chapter X) in one case only. But detailed comparisons of the main group of 127 men (factories A to E) with the 62 men who were interviewed in factory G and the 8 men who were interviewed in factory F showed that these three sub-groups are closely similar with respect to all four of the characteristics set out in Tables 3 to 6. Conclusions in the following chapters can therefore be taken to be equivalent with regard to these characteristics even though they depend on different ranges of information.

The case with the interview records of women workers and of bosses is different because, while both these sets of records covered a fairly wide range (cf. Tables 3 to 6), neither set could be relied upon to be nearly as representative as the set of interviews with the working men. For the women interviewed, there are three distinct difficulties. In the first place, only 10 of 85 (as shown in Table 1) were fully interviewed, in the final form of the interview: so that the information available for the great majority is limited in some respects (factory F: 43 women) or in major respects (factory G: 32 women). Secondly, it is clear that the distribution of interviews by factory (cf. Table 1) gives a quite unrepresentative total. In factory F only a substantial proportion of the women workers volunteered to be interviewed: in factories C, E and G the proportion was far below the incidence of women employed in the factory. The third and most important point is related to this: half of the women workers who were interviewed were in factory F, which, as noted above, cannot be regarded as in the least typical of its industry. The conclusion is that only a very limited use can be made of the interview records of women workers; and these records are therefore referred to in chapter X only, and only for the purpose of making a few summary comparisons with the men whose records are considered in the intervening chapters. For the bosses who were fully interviewed the case is simpler. As noted above, no attempt was made to include a representative sample of bosses of various grades, because this would have been impossible. In consequence, the records were used only for the summary comparisons which are set out in chapter X.

FIELD PROCEDURE AND ANALYSIS

The methods of investigation and analysis require brief discussion in order to clarify their less obvious features, especially those features of the interview procedure which are not defined by the general design of the investigation.

The interview check list (which is given in full in the Appendix) covered five main areas: identification of the person, behaviour at work and opinions of the factory, personal and occupational history, opinions on politics and social classes, and activities outside work, including family responsibilities. The relevance of these areas to the general considerations set out in chapter I is obvious and requires no further discussion. The relevance of the detail within each area was not determinable in advance, as it was not possible to state exact and detailed hypotheses. In consequence, it was found that some of the information gathered had no bearing on the purpose of the investigation. In particular, it was found that too much attention was paid to general ideas and opinions about the work situation and too little to the person's history, family responsibilities and general activities outside work. This was true even for the interview check list used in 1952. In 1950, the check list used was necessarily concentrated on the work situation. The 1951 check list was similar to that used in 1952, only more detailed. These earlier lists are not reproduced: the general point is sufficient, that in all cases an exact and detailed hypothesis would have included more behaviour (overt or covert) outside work and less detail in the work situation.

Much the same applies to the check lists used for interviewing supervisors and executives. For supervisors or professional workers who were willing to be fully interviewed, the check list was as for the workers (that is, as shown in the Appendix), but expanded in the second area to include more detail of their behaviour at work, particularly of their methods of controlling their subordinates. For the persons who were interviewed only as expert consultants (all executives and managers, and some supervisors) the check list was much shorter altogether, and was concentrated on the general problems of factory management, including production as well as labour problems. These check lists too need not be detailed. The same general point may be made, that both derived from a relatively inexact set of hypotheses and both, in consequence, were somewhat more directed to behaviour at work than was required by the interview situation.

However, the form of the interview depended on other considerations than that of the relative vagueness of the initial hypotheses. The instructions to interviewers were that the check list was to be used strictly as a check list; as a list of topics on which information was required. Questions were not to be asked in a fixed order; topics were to be pursued as the person interviewed brought them up himself. Further, questions were not even to be asked in a fixed form of words, but in whatever form seemed appropriate to the interviewer in the particular context. The only exception to these general instructions were the questions on social classes and political opinions (topics 34 to 38, in the Appendix) for which the experience of other investigators (notably Oeser and Hammond (1954)) had shown that an invariable order of questions and form of words was desirable. Apart from this, interviewers were not merely permitted but positively required to adopt a flexible, literary or artistic, approach and not a rigorous or scientific procedure. The method prescribed was probably the more appropriate for gathering the general and rather undefined information that might be relevant to the broad hypotheses adopted, but there were two more important reasons for using this method.

The lesser of these was the consideration of the time required for interviewing. In practice, this was about an hour and a half on the average for each person, with a range from slightly under an hour to over three hours. This considerable interruption of the person's work (and therefore of production) was rather more than the most that the factory managements found tolerable. A more detailed and less flexible interview procedure would have taken considerably more time and would hardly have been acceptable. The chief reason, however, for using the most open procedure possible was that this was thought to be the most suitable for the persons interviewed. It is the furthest removed from the test, quiz or questionnaire situation. It allows the person interviewed the most room to follow his own bent informally: to amplify, digress, recur or be irrelevant, as he wishes. Simply, it is more like an ordinary conversation than a psychological test, and was preferred for this reason. If it is accepted that this kind of interview is unreliable because it is unscientific, then the investigation described below was condemned before it started. But the open interview, whatever its weaknesses may be as a prognostic instrument, has not been shown to be unreliable as a means of gathering information about the person's present. On the contrary, it is probably the most generally useful means (on which see the author's argument (1957, ch. IX)). What is more, it is probably the only method of enquiry which appears reasonable to both the investigator and the persons investigated (on which see Roethlisberger and Dickson, (1927, ch. XIV)).

These considerations were borne out in practice. Interviewing was done by a number of persons, all of whom were graduates in psychology: in the years 1950, 1951 and 1952, by ten, two and three persons besides the author. The prescribed conditions of interviewing were not easy to meet in any year, especially as the interviewer was required to record the conversation on the spot, as nearly word for word as he could. None the less, there were very few obviously defective records and the few that were returned derived, in most cases, from the inability of the person interviewed to state his views, either because of insufficient command of a common language ('New Australians' or foreigners, who were interviewed in German or French when necessary), or because of his having no common frame of reference (a very few extreme neurotics whose only theme was their particular complaints). On the other hand, the conditions of the interview were supposed to be easy to satisfy for the persons interviewed. That this was so is shown by the facts that nobody declined to be interviewed (although some persons were not genuine volunteers), that nobody terminated the interview once it was under way, and that two or three persons only showed some reluctance to speak of what they regarded as their particularly private affairs. Whether a more detailed and rigorous procedure would have been equally acceptable, may be doubted. The other point remains that, in the circumstances of the investigation, a more rigorous procedure would probably have produced less relevant information.

Observation, as a method of investigation, can be disposed of more briefly. It has already been noted that rigorous and systematic observation of behaviour at work would have been impossible, both because of lack of time and because of the suspicions and misgivings of the persons who should have been observed. It must be noted, however, that incidental observation of the performance of actual tasks was possible to a degree sufficient to clarify differences between tasks. Observation of social behaviour, on duty or off duty, was not possible to a useful extent.

The last point that requires comment is the methods of analysis, as set out in detail in chapters III to IX. The reason for comment here is that in some respects these are not conventional methods. In the conventional analysis, both variables and categories of variables are decided before the data are collected or, at the least, before analysis is begun. After that, the variables are reduced to one-dimensional scales, which may or may not be compounded together. The extreme and not very likely case would be that the open interview, as described above, is recorded directly in reduced (or abstracted) form on rating scales. The reduction of interview material to a set of one-dimensional scales, however, is usual. The author has argued elsewhere (1957, chs. IX, X) that the definition of variables and their categories before the event, and the definition of scalar categories, is not necessarily the most useful form of analysis. In the case taken, these definitions were not possible at an early stage, because no exact and detailed hypotheses were possible (or necessary), and did not seem necessary at later stages. The interview material necessarily specified both variables and categories of these to some extent. Beyond this minimum specification, the analysis went on to the delimitation of variables and categories by inspection, without regard to 'dimensionality'. Dimensional or scalable variables were used only where they seemed appropriate, and the greater part of the analysis was therefore of non-parametric associations, depending only on the computation of chi-square. Within these terms, the analysis was undertaken in three stages. The person's considered final evaluations of his work situation (or his reported satisfactions in work) were analysed first. Then the variety of actual work or personal situations was analysed, and related to (associated with) the overall evaluations. Lastly, the person's general situation and opinions were analysed, in relation to his evaluation of his situation at work.

SUMMARY

The detailed historical review of how the investigation was carried out may be reduced to a summary of the main points made. Firstly, the investigation was, for the most part, introduced purely as a research project. An elaborate introductory procedure was therefore necessary, which was on the whole effective in getting the co-operation of the workers, though its effectiveness varied from place to place. The means of investigation was open interviewing throughout, supplemented by some observation of persons at work (see the Appendix for the interview check list). A number of men and women workers, and some bosses, were interviewed in seven factories. The factories were clearly not representative of the whole range of industry in the area in which they are. The workers interviewed could be said, in some cases, to be typical of their factories in certain obvious respects, such as age and length of service. In other factories, and especially in those where few workers volunteered, this could not be said; and in no case could it be said that the workers interviewed were typical of their factories in less obvious respects. It appears, however, that there was a sufficiently wide range among those interviewed by age, length of service and occupational grade for a comparative analysis to be worth undertaking, even though this is necessarily restricted to the numbers of complete interview records that are noted in the following chapters. Lastly, the data were treated by the conservative method of content-analysis, without regard to the conventions of 'dimensionality'.

An important point may be noted in conclusion. The scope, nature and limitations of the investigation should be sufficiently clear. It has not been sufficiently stressed, however, that the investigation was undertaken at a particular time and in a particular place. The conditions of industry in a part of Australia in 1950–2 are not necessarily the same as conditions in other countries and at other times. On the contrary, three particularly Australian features are obvious. There is an excess (compared to other countries) of recent immigrants (in the case taken, mostly British, but some continental Europeans). Besides this, it is generally accepted that the legal definition of hours of work and wages favours the working man (40 hours a week and a relatively high guaranteed minimum wage). Thirdly, there are doubtless features in the Australian outlook on social classes, as well as in the Australian way of life generally, that do not have a direct counterpart elsewhere. Besides these local factors, the conditions in time of the investigation may be important. In Australia in 1952 there was almost full employment: in 1951 there was full employment except in two or three industries; and in 1950 there were vacancies in all industries. That the findings set out in the following chapters would hold in other conditions (that is, in the conditions of extreme under-employment of the past or hypothetical future) is improbable.

CHAPTER III

THE WORKER'S EVALUATION OF WORK

THE first part of the general analysis is the summary of the statements made by male factory workers about the various aspects of their work situation. In the course of the interview, each worker stated how he liked his job, what he thought of his pay, and how he got on with his bosses of various grades and with his workmates. The comments made under these four heads varied widely, so that records of identically worded statements were rare. The summary classifications set out below are therefore a practical compromise between categories the majority of which would include one statement only and categories of broad content because they are in the same dimension. The compromise, that is to say, is the final reduction of a more complex analysis, and is intended to hold the balance between the chaotic proliferation of detail on the one hand and the unmeaningful suppression of information on the other.

SATISFACTION IN WORK

The four distinct aspects of the whole work situation are taken in turn, starting with the worker's comments on his actual task, as shown in Table 7. It is clear that these are not made in one dimension, or in terms of one factor only. The worker was asked his opinion of his job, not specifically of his task, or the actual

TABLE 7

Workers' comments on their tasks

The task is	Frequency	Per cent
1. Varied, interesting: good conditions, amenities may also be mentioned	36	19
2. Easy, has good conditions, amenities: good pay or boss may also be mentioned	47	25
3. Excellent, very good, good, O.K., no further detail	48	26
4. Not mentioned: the *job* is good because of good pay, bosses, etc.	14	7
5. Dull, dirty, hard, tedious, boring: bosses, pay, etc., may be mentioned, as good or bad	44	23
Total	189	100

work he did. Some workers referred at once to the whole work situation, including bosses, pay, mates and possibly other elements. Even if the variety thus introduced into the responses is disregarded, there are three distinct kinds of good task: those that are interesting (category 1), those that are easy (category 2) and those that are simply 'good' unqualified and unexplained (category 3). It should be noted that remarks were placed in this last category only when it was clear, by the elimination of all other comments on specific elements in the work situation, that this one brief comment was all the worker had to say about his task. As this was his spontaneous opinion, it was accepted without further enquiry. In the same way, it was accepted without further enquiry that, if the worker did not say even one word about his task, this too was his spontaneous opinion. In consequence, there are some workers to whom the task itself is apparently quite unimportant (category 4) and a much larger number who find the task objectionable in various ways (category 5). It is just arguable none the less that the categories, as they are set out, are in rank order. For comparative purposes, it is safer to group the first three and the last two categories (as further analyses depended on the use of dichotomous judgements of the work situation) and to conclude that 70 per cent of the men were more or less satisfied with their tasks and 30 per cent were not satisfied. For this dichotomy, there are no differences associated with age, length of service, nationality or occupational grade (i.e. tradesmen, process workers and labourers). These findings are generally in line with those reported by Viteles (1954, pp. 9, 10) if account is taken of the fact that only 23 (not 30) per cent of the men interviewed are actually dissatisfied with their tasks. More detailed comparisons are hardly possible. The interesting figures presented by Walker and Marriott (1951), for example, include relatively more men who find their tasks interesting and also relatively more who complain of boredom or bad conditions of work. These differences are due in part to differences of method and in part to the range of jobs studied. It may be possible to reach some conclusions about the conditions under which a task is regarded as interesting, boring or difficult, but this question is best left till chapter IV.

The workers' opinions of their pay are set out in Table 8. This set of comments is perhaps not far from being a simple scale, as category 2 clearly denotes less satisfaction than category 1, and categories 3 and 4 denote dissatisfaction. However, the order of these last two is not clear, although the difference in their content is. It is sufficient to take the obvious dichotomy between categories 2 and 3 and to note that 66 per cent of the workers were more or less satisfied with their pay. For this division, there are no

TABLE 8

Workers' comments on their pay

The pay is	Frequency	Per cent
1. Terrific, excellent, too much, very good, good	64	34
2. Good, sufficient, considering: job is inferior *or* can make extra (overtime, etc.)	61	32
3. Not enough to live on: may also be called not bad for the job	35	19
4. Not enough for the job, which is skilled, unpleasant, better paid elsewhere, etc.	29	15
Total	189	100

clearly significant differences associated with age, length of service, nationality or occupational grade. No direct comparison with other findings can be attempted, as little or no attention seems to have been paid to workers' simple opinions of their pay, as distinct from their ratings of the importance of pay. The general point may be made, however, that the findings reported obviously depend on the context. To ask: 'What do you think of your pay?' is not the same as to ask: 'Do you think your pay is enough?' or, at the extreme: 'Would you like more pay?' It seems altogether likely that far more than 34 per cent would have answered 'yes' to this last question. It is therefore likely that this percentage is a minimum estimate of the fraction who were actively dissatisfied with their pay. These, like the others who were more or less satisfied, were working in a country which is generally held to have rather high rates of pay, whether absolutely or in relation to the working week. On the other hand, they were interviewed at times of full employment, or even when there were jobs vacant (1950). The judgements made (if they were considered judgements) may therefore have been related, in some cases, to the general economic possibilities.

The analysis of workers' comments on their bosses is a little more complicated, as comments were made on bosses of different grades. Early in the investigation, it was thought that the important distinction would be between the foreman and the top management, comments on whom could be considered separately. This idea proved to be too simple. It was found that few workers had any idea of top management as a policy-making collectivity and this particular point was therefore included in 1950 only (i.e. in factory G only). On the other hand, many workers knew a number of bosses by name and were prepared to give opinions on their personal qualities, efficiency and so on, as the attributes of specified individuals. All workers knew and gave the correct or official rank order of the bosses mentioned: all, that is, who mentioned more than one boss placed the foreman above the leading hand or the manager above the foreman. Beyond this, there were wide individual variations in the reports. Firstly, in the number of grades distinguished; some workers knew only their immediate boss while others knew persons in up to four grades above the lowest. Secondly, and independently, the starting points of the scales used varied. Some workers reported the leading hand as their immediate boss; some the foreman, even when there was a leading hand; some a professionally qualified supervisor (chemists in factories A and B), even though he had foremen under him; and one worker in a particularly rare trade, reported the production manager as his immediate boss. In all of these cases, even the last, the worker might state that he knew no higher boss or that he knew bosses in one or more higher grades, up to managing director. Thirdly, there was variety in the degree of knowledge stated, and this was not uniquely associated with the grade of the boss. The least statement that was regarded as signifying some degree of knowledge was of the vague kind: 'I know him by sight, he looks all right/sour/never says anything', and so on. At the other end of the range were occasional long and detailed accounts of what a particular boss had done over a period of years. The more detailed reports were usually but not always of lower-grade bosses and the vague were usually but not always of higher grade bosses.

It was clear that the variety of reports reflected the considerable variety in work situations (and, for some persons, in length of service too). But what was wanted was the worker's judgement of the situation as he saw it. It was therefore decided to accept the variety presented and to ignore the official hierarchy in so far as some workers had ignored it. Table 9 shows a summary of the comments in which those on the worker's immediate boss are compounded with those on all higher bosses. It is obvious that this set of six categories is no scale. Indeed, it may be doubted that either part is a scale. Comments on the immediate boss include some vaguely favourable, which may none the less be as favourable as the more defined comments; and include at least two distinct kinds of unfavourable comment. With the higher bosses, provision had to be made for their being unknown, but this cannot be regarded as a neutral comment. The unfavourable comments, again (in categories 3 and 6) are of two, if not three, distinct kinds.

These comments had to be treated as one set, for the reasons

TABLE 9

Workers' comments on their bosses

The immediate boss is	The higher bosses are	Frequency	Per cent
1. Extra good, lets alone,	Nice chaps, lovely, gentlemen	64	34
2. helpful, fair, nice chap	Unknown	25	13
3.	Some good, others not; too remote, inefficient; crooks, twisters	46	25
4. Pleasant, but incompetent; unpleasant,	Nice chaps, lovely, gentlemen	13	7
5. crook, quéer, moody,	Unknown	16	8
6. rotten, a pig	Some good, others not; too remote, inefficient; crooks, twisters	25	13
	Total	189	100

outlined above. The only alternative would have been to accept only comments on persons of certain defined grades (e.g. the foreman and the production manager), which would have meant omitting a substantial number of interview records. In any case, the procedure adopted was appropriate to the purpose of the investigation. It may be noted that immediate bosses are significantly better liked than are the higher bosses: the total of categories 1 to 3 in Table 9 is 72 per cent, while the total of categories 1 and 4 is only 41 per cent. But the reverse does not hold for positive dissatisfaction. If categories 4 to 6 are compared with categories 3 and 6, the percentages dissatisfied are 28 and 38, and the difference between these is not significant. Judgements made on the two sets of persons may well differ only because 21 per cent of the higher bosses are unknown. The idea that these judgements therefore represent a coherent set of opinions on all bosses known is further supported by the fact that the two sets are slightly but significantly associated ($P < 0{\cdot}01$, $\phi = 0{\cdot}22$).

In any case, the compound set gives what is wanted for further analysis: a distinction between the workers who think well of all the bosses they know, and all other workers. For this division, between category 1 and categories 2 to 6, there are no differences associated with length of service or nationality. The younger workers (up to age 29) are significantly less satisfied than the older (age 30 and over) and the tradesmen are significantly less satisfied than the process workers or labourers. These differences may be cultural effects: it may be easier for the younger man or for the tradesman to assert his independence by being critical. On the other hand, considered in relation to the number of differences tested and found insignificant, these two (neither of which is marked) may be only sampling accidents.

In this case, as with the workers' comments on pay, no direct comparison with other findings is possible. It may be noted, however, that the final figure of 34 per cent satisfied with all the bosses they know, seems low. It is depressed in part, of course, by the rigours of definition. On the other hand, it derives from seven factories which were probably (with one partial exception) among the more enlightened in their methods of management and control. The implications of this, however, must be left to chapter X.

It remains to consider the workers' comments on their relations with their workmates, which are shown in Table 10. These comments come nearest to being in an obvious and simple rank order,

TABLE 10

Workers' comments on their workmates

The worker's relations are	Frequency	Per cent
1. Excellent, very good, matey, good, all right, no snags	108	57
2. No strife; not bad, but prefers to have little to do with mates	23	12
3. Good, but poor with cliques or individuals (foreigners, old hands, etc.); or poor generally	58	31
Total	189	100

though this would be clearer with a larger set of categories, up to five or six. As it is, the division between satisfactory relations and cautiously neutral or unsatisfactory is between category 1 and the others. For this division, there are no differences associated with age, length of service, nationality or occupational grade. It may be surprising that only a slight majority of workers report positively satisfactory relations with their mates. In part, however, this may only reflect the relative unimportance of these relations in the work situation; hence 12 per cent in category 2. It should be noted also that very few persons in category 3 (6 of 58, or 3 per cent of the total) reported actively hostile relations in general.

A conclusion on satisfactions in work can now be suggested. As the investigation was carried out in relatively enlightened, superior or intelligently managed factories, it is interesting that the percentage rates of positive satisfaction are not high for any of the four elements of the work situation that have been distinguished. The point cannot be pressed. Besides the obvious qualification that what one means by high or low depends on the context, there is the difficulty that direct comparisons of the elements of job satisfaction with those reported in other findings is not possible in detail because other investigations are not, for the most part, in comparable terms. The elucidation of the worker's satisfactions in work by the analysis of relatively undefined comments, which are for the most part in his own vocabulary, is likely to give results which differ from those of the usual kind of morale survey, if only in that they are more detailed. At all events, it can be taken that the first stage of analysis has produced an intelligible set of comments on four distinguishable aspects of, or elements in, the work situation as a whole. It can be taken also that, within the general limits of the investigation, these comments are reasonably representative for workers who have settled down in their jobs; that is, that the very few sociologically linked variations noted above can be disregarded. The persons investigated can therefore be treated as a homogeneous group, though not a general industrial sample (cf. chapter II). The next step, therefore, is to explain the comments made by analysing their derivation. This analysis (except only for certain findings in chapter X) is based on the dichotomous reductions of the sets of comments that were noted above; a more complex analysis would have been statistically unmanageable.

Before the analysis can be undertaken, however, one must consider if it would be worth doing; and this would not be so if the most that can be said about the summarized comments is that they are an intelligible set of opinions on aspects of the work situation. They may be deliberate and articulate judgements of this situation, or only improvised and unconsidered opinions given in the interview situation. They may be all of importance as deliberate judgements which determine or lead on to definable actions in the work situation, or they may differ in importance in this respect. Lastly and most obviously, the comments analysed may signify distinct satisfactions in the work situation, or merely related aspects of satisfaction in general. The evidence on this last point may be considered first.

The appropriate test is the associations among the four summarized satisfactions in work, when each is taken as a simple dichotomy with the divisions noted above. It was found that these associations were all statistically insignificant for the whole population of 189 men. Subsidiary findings were that satisfaction with the task and with pay are significantly associated among younger workers (up to 29) and among tradesmen, and are negatively associated among the oldest workers (age 40 and over). These associations are of little practical significance ($P < 0{\cdot}05$, $> 0{\cdot}02$ in all three cases). It can be concluded that, for all practical purposes, the four satisfactions in work are independent. A consequence is that all of the sixteen possible combinations of these satisfactions (arranged as dichotomies) were found, as shown in Table 11. The further analysis, of course, was not directed to this complicated set of categories but to the independent and simple dichotomies described above.

It may be concluded that the finding of independence is, on the whole, to be expected. In the first place, it is not likely that the four different aspects of the work situation would be found to be associated if they were assessed by a detached observer. There is

TABLE 11

Combined satisfactions in work

The worker is satisfied with his	Frequency	Per cent
1. Task, pay, bosses and mates	23	12
2. Task, pay and bosses	7	4
3. Task, pay and mates	36	19
4. Task, bosses and mates	7	4
5. Pay, bosses and mates	4	2
6. Task and pay	26	14
7. Task and bosses	7	4
8. Task and mates	11	6
9. Pay and bosses	9	5
10. Pay and mates	10	5
11. Bosses and mates	5	3
12. Task only	14	7
13. Pay only	10	5
14. Bosses only	2	1
15. Mates only	12	6
16. Nothing at all	6	3
Total	189	100

nothing in the nature of industry that would tend to make the easiest or most interesting tasks the best paid also, or vice versa. There could be associations due to management's initiative, or lack of initiative. Thus it might be expected that a management which tries hard for good labour relations will also be inclined to improve conditions of work and amenities, and perhaps to pay more than other factories; and vice versa. It seems unlikely, however, that management's initiative could affect relations among workers, and the effects suggested are likely to be limited. Whether a task is interesting or dull, and even whether it is easy or hard, are largely given in the nature of the task, and are not that much open to alteration, short of radically re-designing the task. Rates of pay, again, are largely defined by legal procedures for whole industries, and there is limited scope for variation above these definitions. Finally, associations of this kind would be likely to appear only in data drawn from a wider range of situations than in the case taken. Besides that all the factories investigated were relatively progressive, the investigation was at a time when it would not have paid any management to try to maintain the bare minimum in labour relations, task conditions or pay.

A second possibility is the case which was not found, that satisfactions in work are associated because this is how the worker sees them. This is not necessarily a simple 'halo' error; it is conceivable that one considered judgement might influence another. If two workers, for example, get the same pay for doing different tasks it is conceivable that the one whose task is hard and dull would make the uninfluenced comment that his pay is poor, while the other, whose task is interesting and easy would make the considered judgement that his pay is good. However, the conditions under which such a transfer might occur are unlikely to be found. It is possible to imagine the worker who likes his task going so far as to say his pay is not really enough, but he is not really sore about it because he likes his work. It is more difficult to imagine him saying his pay is all right, let alone good, simply in consequence of his liking his task. This kind of transformation of his perception of the situation would occur, presumably, if he were burning with enthusiasm for the task; if he were so deeply involved in the task that he suspended judgement on other aspects of the work situation and, in effect, made the one judgement. Involvement of this degree is, on the face of it, not likely. The converse case, that the person so dislikes one aspect of his situation that he makes one negative judgement only (cf. Viteles (1954), p. 307), is perhaps easier to imagine but again is not likely to occur.

The third possibility is that of the simple 'halo' error. The person makes the same judgements, positive or negative, of aspects of the work situation which, in the interview setting, he thinks ought to go together, even though it is not necessarily his considered judgement that they do go together. This apparently did not happen. As this sort of error is obviously more likely with opinions manufactured on the spot than with considered judgements, the independence of the four satisfactions in work may be taken as some indication that they were indeed considered judgements.

THE NATURE OF THE WORKER'S JUDGEMENTS

If these are considered judgements, they must have consequences in behaviour; so it can be shown that they are considered judgements by showing that they do have consequences in behaviour.

The difficulty here is that direct evidence of behaviour at work is lacking, and indirect evidence is inconclusive. Some use may be made, however, of the reasons for leaving previous jobs that were given by the 127 men interviewed in factories A to E. These reasons were usually stated in some detail, and were of several kinds. According to his present report, the person may have left his job because it was finished or because he was dismissed when the firm was wound up. Or he may have left for family reasons; because he wished to return to the city, to be near his parents, and so on. Or, lastly, he may have left because he was dissatisfied with some continuing element or elements in the work situation; he thought the job was insecure or had no prospects, that his task was unnecessarily dirty, his pay low and likely to go lower, that his bosses were unfair and given to bullying or that his mates were inclined to sneak to the foremen. The data for dissatisfactions with the relevant four elements in the work situation are summarized in Table 12.

One conclusion is immediately apparent. The person's dissatisfactions, as shown in row 3 of the table, were the effective grounds

TABLE 12

Reports on past and present work situations

	Task	Pay	Bosses	Mates
1. Past: number who changed work because dissatisfied with	76	54	49	11
2. Past: number who report no such action	51	73	78	116
3. Past: percentage actively dissatisfied	60	43	39	9
4. Past: total number of times work was changed	143	75	80	12
5. Past: average number of times per person (row 1 only)	1·9	1·4	1·6	1·1
6. Present: number dissatisfied with	29	39	69	44
7. Present: number satisfied with	98	88	58	83
8. Present: percentage dissatisfied	23	31	54	35
9. Ratio of past to present dissatisfaction (row 3/row 8)	2·6	1·4	0·7	0·3
10. Ratio of past to present dissatisfaction (row 4/row 6)	4·9	1·9	1·2	0·3

of his action, in the limiting case that he simply went elsewhere. It is not possible to press this conclusion much further. Obviously no inference be drawn about the consequences of dissatisfaction in daily behaviour at work, because the latter is unknown. Further, the dissatisfactions reported are not directly comparable to those that were reviewed above. Their detailed content, as may be seen from the brief examples given, is not dissimilar. It must be assumed, however, that, so far as the person's report can be relied on, what he reported was an extreme degree of dissatisfaction; a positive dislike of the task, pay, bosses or mates that was as marked as, if not more marked than, the greatest dislike reported in the present situation. Even so, the limiting case stands: the considered judgement that an element or a combination of elements in the work situation is objectionable, may lead directly to action. It can therefore be assumed that the person's judgements of his present situation are equally important as potential grounds for action even though their consequences cannot be demonstrated for lack of evidence. Lastly, the point is worth making that the reasons given for leaving previous jobs, in so far as they were not invented in the interview (which is unlikely), were certainly considered judgements. To leave a job is no light matter for anyone, especially for a manual worker, and it is inconceivable that it would usually be done without full and careful consideration of the consequences in relation to the unpleasant features of the present situation. This makes it the more likely that the reports on the worker's present situation are also considered judgements and not merely casual opinions.

A second conclusion can be drawn from the data. It would seem that the four dissatisfactions in work were not equally effective as grounds of action, but that they could be ranked in order of effectiveness following the percentages shown in row 3 of Table 12. The appropriate tests are of the pairs or figures in rows 1 and 2 of this table, and it is found that all differences between pairs are significant except for the difference between pay and bosses. The provisional conclusion is that the rank order of effectiveness (or importance) is task, then pay and bosses equal, then mates. However, this takes no account of the number of times the person took action on any particular ground and in this too the grounds differ, as shown in row 5. The appropriate test in this case is of successive pairs of figures in rows 1 and 4, and only one significant difference appears, that between task and pay. The incidence of action only reflects the number of persons acting on each ground.

But these tests depend on the assumption that the persons in this sub-population of 127 were equally exposed to objectionable tasks, pay, bosses and mates. If more persons left their previous jobs because the task itself was unpleasant than left because the pay was too low, this may show that the unpleasant task is the more important ground for action. Equally, it may be due only to the fact that unpleasant tasks are encountered more frequently than low pay. There is no direct evidence of the incidence of objectionable elements in past work situations, as the person's report of his main reason for leaving any previous job is not comparable to the detailed assessment which he might have made, if he had been asked, of each previous work situation. However, equal incidence is unlikely. Another possibility, as noted above, is that the incidence of objectionable features is the same as the incidence of action, so that there is no order of importance among the elements of the work situation. The third possibility is that incidence is as it is reported in the present situation. Clearly it is so and, with one qualification, it is the likeliest case that the present incidence fairly directly reflects the past. The industrial situation has changed, of course, since the times that the men concerned left various of their previous jobs, which might have been as long as 45 years before, or only a month before, they were interviewed. However, as far as satisfactions in work are concerned, all changes are likely to have been in the same direction. The qualification necessary is that, if the past is to be represented by the present, it should be by positive dissatisfactions in the present; the person objects to his task or his bosses, he is not merely indifferent to them. In Table 12, therefore, the entries in row 6 are in three columns derived from fewer categories than those included in the working dichotomies discussed previously. For the task, this row includes category 5 only; for the bosses, categories 3 to 6 but not category 2 (which could hardly be a ground for leaving one's job); and for workmates, category 3 only.

For this hypothetical basis of the incidence of objectionable elements in the work situation, two sets of tests between successive pairs of figures are possible. Present dissatisfactions (rows 6 and 7) may be compared directly with past (rows 1 and 2). In all four cases, it is found that the differences are significant, and their direction is shown in row 9 of Table 12. Thus far more persons left previous jobs because the task was objectionable than were hypothetically exposed to objectionable tasks, and so across the column. The order of effectiveness in determining action now has four distinct ranks: task, pay, bosses and mates, in order. The second set of tests is the comparison of the number of times action was taken (row 4) with the hypothetical incidence of objectionable features (row 6), for each pair of elements in turn. All six differences are significant, five of them highly so ($P < 0{\cdot}001$), and their direction, as shown in row 10 of the table, is the same as before. Compared to the hypothetical degree of exposure, objections to the task were a powerful ground of action, objections to workmates a feeble ground, and so in between. The first conclusion was that the workers' comments on their present situation are probably considered judgements that would be effective in promoting action even though that cannot be defined in detail. The second conclusion can now be stated: that, in the limiting case of leaving the situation (and therefore possibly in other cases for which there is no evidence) there is a clear and distinct order of importance or effectiveness in promoting action. The task matters most, then pay, then the bosses, and workmates last and least.

This second conclusion may be compared with the findings from other investigations, subject to the usual difficulty of finding a basis for comparisons. There is no lack of opinions about what the worker really wants or, by implication, acts on. Two contrasted examples of these will suffice. Gardner and Moore (1950, p. 227) observe that: 'The typical factory worker is far more concerned with his relations with his fellow employees than with earning an extra dollar or two under a piecework system.' This may be true. There is nothing in the present investigation to contradict it, as very few men in Australia work under piecework systems, and none of those interviewed did so. It does raise the question, however, of the conditions under which the worker's relations with his mates are a primary concern. Against the tendency in the more recent industrial psychology (of which the above is merely an example) to assume that social relations are important is the observed fact that, in some circumstances at least, they are quite unimportant. Mace (1948) more cautiously notes that satisfactions in work may depend on the relations between work and the worker's other activities, but that some interest in the task itself is probably necessary for satisfaction. These views are developed no more than is possible within the limits of a brief essay but they have, at the least, the merit of being intelligent. That work may be satisfying because of the remoter purposes it serves is a neglected point that will be taken up in later chapters. The point that matters here is that what the worker is paid to do, and what he spends nearly all of his working time doing (unless he wishes to invite dismissal), is his task. There may be circumstances in which the social relations incidental to this, or even additional to it, are a primary or compensatory satisfaction. To suppose that these are the usual circumstances is to leap ahead of the evidence.

Unfortunately, the evidence is scanty and some of it is ambiguous. In the latter class fall two extensive surveys discussed by Viteles (1954, ch. XIII). It is a little difficult to relate the results of these to the four relevant satisfactions in work, but a rough classification of content gives a rank order for these, in importance to the worker, which somewhat resembles the one reported above: task, bosses, pay and mates. More important than this very vague confirmation, however, is the consideration of the basis of the findings. In the lesser survey, which included only some thousands of workers, the worker selected the 5 most important of 71 listed 'morale factors'. His judgements were relatively contextless, as they were not directed even to hypothetical action, so that comparable results are hardly to be expected. On the other hand, the judgements made were limited to the context of the list presented, which probably accounts for the fact that relations with workmates hardly appear at all in the final percentages. In the greater survey, a sixth of a million workers wrote letters to describe what they liked about their jobs, as part of a competition arranged by the corporation which employed them, and which offered over 5,000 substantial prizes for the 'best' letters, as judged by an independent panel. It is not surprising that the results ot this survey, if it can be taken as a survey, are not comparable to those discussed above. In particular, relations with the bosses clearly take the first rank, which, in the circumstances, is not remarkable. The only conclusion that can be drawn from either of these massive efforts is that, if one wants to know what are the grounds of the worker's behaviour, it is more important to ask directly relevant questions than to investigate enormous numbers of persons.

More directly comparable findings appear in a report of Katz and Kahn (1952), who asked an unspecified group of workers to state the things they liked best about their jobs. Here again, there is no reference to action, but the procedure was open: the person stated, and did not merely vote for, his preferences. The results reported may be summarized to give this rank order: task first, pay and bosses equal second, mates last. The most direct comparison that can be made is with the findings of Reynolds and Shister (1949), who asked about 800 American factory workers why they had left their previous jobs. Their results give exactly the same rank order as that of the present investigation (task, pay, bosses, mates) and the differences between these four elements are all highly significant.

It may be concluded generally that the satisfactions in work set out in this chapter should be capable of explanation by further analysis. It seems probable that they are, to some extent, considered and deliberate judgements. Some may have been vague guesses made on the spot. Some may have been made for the first time with full awareness in the interview, but need be none the less considered and deliberate for that; and some were probably made openly in the course of the person's ordinary work. Finally, the judgements made can be related to action in certain limiting circumstances, and to different degrees; so that they could probably be related to action in more general circumstances.

However, it is not suggested that these four judgements give a complete account of the work situation as the worker sees it. Obviously, they derive from a classification of the interview material which is to some extent arbitrary. It would have been quite possible to include all four judgements under one head, as a 'morale scale', in spite of their observed independence. It would have been equally possible to break up each into components, or separate judgements of aspects of the task, the pay system, and so on. Again, it would have been possible to include material that was not used in the definition of judgements or even, by using a longer check list, to include different material altogether. In this last case, it would have been possible to consider judgements of major aspects of the work situation, such as security, that were simply not included. Thus there are two questions to be considered: the restrictions of detail imposed chiefly by the method of analysis, though to some extent also by the nature of the interview check list, and the major restrictions that may have been introduced by the omission of certain topics.

The short answer to the first question is that time did not permit either the use of a long and very detailed check list or the analysis of every detail that was none the less elicited by the interview. Thus there was no particular enquiry into such job hazards as dust, grease, smoke, fumes, steam, heat, noise, ventilation, fleas, flies or rats. Where any such details were mentioned by the worker of his own accord, they were classified together under the one head of objectionable features of the task. However, this was not done only for convenience. The prior assumption was made that judgements that are likely to have continuing (that is, stable or recurrent) consequences in behaviour are likely to be judgements of continuing general aspects of the work situation rather than of small, probably unrelated and possibly temporary, aspects. Although this assumption cannot be tested directly, it seems probable, as can be shown by the consideration of some of the kinds of behaviour that would have been relevant to a test.

With the one kind of behaviour that has already been discussed, the worker's staying in his job or leaving it, there is little difficulty in the assumption. He is not likely to leave his job, a considerable undertaking in any circumstances, merely because his task is dusty or because there are fleas in the workshop, or because his boss has occasionally looked at him sourly. If enough such small things occur regularly, he may form the considered judgement that his task is objectionable or his boss hopeless, and then leave. Short of this, he is likelier to regard the things mentioned as inconsiderable trifles, if indeed he is aware of them at all until he is particularly asked about them. The converse is equally likely. If the worker has other weighty reasons for changing his job, he is not likely to remain merely because his task is not dusty, or even because he knows it is exceptionally clean.

However, this is behaviour in conflict or under some stress, and the case with ordinary daily behaviour at work may not be so obvious. At the appropriate level of detail, it is clear that occasional events do affect behaviour. If there are fleas in the workshop, the worker kills those that he can catch. Further, relatively trivial events that happen often enough to be noticed as recurrent and then comprehended in a judgement are likely to lead to recurrent action. The worker prepares to attack the fleas or to scowl back at the boss. It cannot be assumed that behaviour at this level of detail, and therefore the judgements from which it derives, are unimportant merely because they were necessarily outside the scope of the investigation. None the less, it would seem more profitable in principle to investigate more general modes of behaviour, such as the worker's recurrent relations with his boss or general task performance, described as one set of interdependent behaviours. It is not likely that behaviour at this level of complexity would depend on the isolated judgements of many unrelated events. On the contrary, abstract and summary judgements like those already discussed would probably be more easily relatable to general modes of action.

There remains the awkward case of the work situation in which the action which the person's judgements point to is blocked by other judgements: the conflict situation, that is, in which action is hardly possible. The person may hate his task and his boss and regard his pay with contempt and his mates with weary indifference. If his judgement of the economic situation is that he would certainly ruin himself by leaving his job, he stays. Of course there are consequences in his daily behaviour at work of this conflict between two sets of judgements. For example, it is usually possible for a disgruntled worker to find subtle ways of holding up the task and obstructing the boss. But these will not be the only consequences. In principle, one should consider also the person's covert behaviour, such as his waking fantasies and nightmares. These are none the less behaviour for being covert. They are not purposive behaviour because the person cannot propose or intend to act in the situation. So what he cannot do in reality he does in fantasy, in his inarticulate or sub-intentional behaviour. This occurs, of course, in all situations, not only in conflicts. In the latter, however, it may be important as an intervening but partly independent cause of the person's daily behaviour at work. The more he is annoyed by the obvious ineffectiveness of his aggressive fantasies, in addition to his frustration by the situation itself, the likelier he is to find the means of subtly handling the situation in his own way. There is no question of trying to study sub-intentional behaviour at work, for the reconstruction of this, as the author has suggested elsewhere (1957, ch. IX) is bound to be even more tedious than the reconstruction of deliberate intentional judgements. The point that matters here is that the kind of behaviour described would obviously not derive from judgements of trivial events, whether recurrent or not. If the worker hates the whole of his work situation so much that he has nightmares about it, this (for the normal person) can only be because he has reached the considered judgement that his situation is hopeless in all respects at once.

It may be concluded that the loss of detail due either to its omission from the interview or to its suppression in analysis was not important; or, indeed, that the suppression of detail in the interest of reconstructing general judgements was the correct procedure. This is a speculative conclusion, which can be only partly justified by the success, such as it was, of the subsequent analysis of these judgements. So little work has been done on the elementary classification of behaviour that, for the most part, what behaviour is recurrent and what behaviours go together are anyone's guess.

There remains the more difficult question, whether the complete omission of certain possibly important topics was reasonable. Attention has been given to judgements made of the most obvious elements in the work situation: the worker's physical and social relations at work, together with pay as the necessary means of entering relations outside work. It may be that other elements, not merely detailed parts of these four, but elements of different kinds, could and should have been investigated. The four judgements set out certainly do not bear a very close resemblance to the usual lists of the various needs which are thought to motivate the worker. The chief reason for this is that the investigation was intended to elucidate the worker's judgements of simple and definable situations, and not his inferred needs or general action-tendencies at a higher level of abstraction. It would be pointless to argue whether or not something has been lost by concentration on the reconstruction of relatively simple and definable covert behaviour and its antecedents and consequences. That question can only be settled in practice, by the results of many years of investigation, not of one limited project only. It is relevant, however, to observe that the contrast between abstract general needs and specific judgements may be more apparent than real. Some of the worker's more general needs can easily be reduced by analysis to components which either were included in the investigation or were excluded for sufficient practical reasons. To argue this for the vast variety of needs the worker has been alleged to have would be tedious; it will suffice to take four cases that are common in industrial psychology.

The simplest of the four is advancement or the prospect of

advancement, which appears prominently in the job-satisfaction surveys summarized by Viteles (1954, ch. XIII). This clearly could have been investigated directly in the interview, as it was by the different methods of selecting morale items and of competitive essay-writing that were used in the surveys referred to. In the interview, it would have been necessary only to ask the person if he thought his job offered any prospects of promotion to foreman, or any prospects of his increasing his knowledge so that he could move on to a better paid job; which is a second and subsidiary, but not unimportant, meaning of advancement. The direct enquiry was not made in the main investigation as it was thought that in Australian factories the prospects of advancement would be of little interest. This depended partly on general experience, but also on the experience of asking the relevant questions in factory F only. The decision not to go on with these questions was fully supported by indirect evidence from factories A to E. Among the requirements of a good job, as stated by the worker (data not referred to elsewhere), prospects of promotion come a poor last: the task and its conditions, pay, bosses and mates were each mentioned by an average of 48 per cent and prospects of promotion by none at all. The possibility of promotion was mentioned by only one worker as one of his present job satisfactions, and two (of 127) workers mentioned that they might leave their present job for advancement; one, to be a boss and the other, to get a better paid job. Finally, 7 persons reported that they had left previous jobs because of the poor prospects of advancement, and some had done so several times, so that the total number of occasions was 23. On both counts, this dissatisfaction would rank ahead of objections to workmates as a ground for action. However, these counts include both senses of advancement: promotion, which had been unsuccessfully sought or obtained but not held, and also getting a better paid job (e.g. a more complex job in the worker's own trade), which had been done successfully. The importance of either sense alone is evidently low. Besides this, it is clear from the other findings that the settled worker accepts the fact that there is little scope for promotion or for improving his job. In either sense, the possibility of advancement hardly concerns the Australian worker, so that it was properly disregarded.

The need for achievement may be considered next. However this term may have been defined previously, it seems to be capable of only two precise meanings in the work situation, both in relation to the task. The worker may feel that he achieves something worthwhile if he does a varied and interesting task. But this is more easily elicited (as it was) by asking for his general opinion of his task than by asking him to judge the task in the abstract terms of trait psychology. Or, more narrowly, the worker may feel that he achieves something in his daily routine if his task is relatively complete; if he makes a whole product himself and can see the finished article. Here again, the enquiry in abstract terms is not necessary, nor likely to be profitable. It is sufficient to note whether workers who are engaged on tasks of this kind (on which see chapter IV) are more satisfied with their tasks than are other workers.

Much the same goes for the equally vague and broad need for recognition, some meanings of which are discussed by Viteles (1954, chs. IV and XVII). In the work situation, this may mean that the worker wants his work to be recognized by his being paid well or that he wants to be recognized as a person (to be spoken to politely, have his efforts and qualities personally acknowledged) by his bosses and mates. All of these are elicited, without any direct enquiry in abstract terms, by asking for general opinions of the relevant factors. It is true that the final summaries of the worker's four judgements confound the comments that directly bear on recognition in any of these senses with other comments. However, this is merely another case of ignoring excessive detail; and that it would be excessive is shown by the rareness of the specific comments so confounded. Another possibility is that recognition (or achievement) refers to events outside the work situation altogether. The person wants to be recognized as a worthy citizen by his family, friends and neighbours; he wants the social status which derives from a highly esteemed occupation (cf. Viteles (1954, ch. XVII)). This might or might not be of importance, but it is so abstract and complex an idea that direct enquiry would hardly be possible. The likelier way to determine whether the person judges his occupation and his general place in society as being in or out of line is to ask him a number of questions about the structure of society and his place in it, to reconstruct his judgement, if possible, and then to relate this to his judgements of the work situation. These matters are set out in detail in chapter VII: the point suffices here that, whether or not they are important, their omission would not leave a gap in the description of work as the worker sees it because they are not elements of the work situation at all.

The last, most important and most difficult example that may be considered is the need for security (cf. Viteles (1954, ch. XVI) and many other discussions). As a possible judgement of the work situation, this has at least three distinct senses. The worker may judge that he is secure from arbitrary dismissal: the boss is not going to sack him for a trivial reason or no reason at all. As this judgement is clearly implied, in principle, in the worker's general opinion of his bosses, it requires little discussion. It may be noted that the specific judgement was never made; among other probable reasons, because arbitrary dismissal at the time of the investigation would have been expensive to any management. The equivalent judgements did occur, however, among the reported reasons for leaving previous jobs: the person left because dismissal for no relevant reason was generally not uncommon, or was probable, or actually happened. In some circumstances security in this sense might well be important enough to warrant separation from the general judgement of the bosses. Had this been necessary in the circumstances of the investigation, it could have been done.

In this sense, security is merely an element, and probably a minor element, of the person's present situation at work. The other senses of the term are more general, as they refer to the hypothetical future. In the present situation, the person has relations with things and with other persons; and his work provides the necessary means for his sustaining relations outside work. He judges his present relations at work, and his rewards for working, as if the present situation is obviously stable and permanent. But this point has not yet been considered. The work situation, however the person judges its present characteristics, may be likely to go on as it is now, or it may have an obscure and unpredictable future, or it may be likely to break down. It may be difficult to take into account this quite different dimension of the work situation, but it is important that it should be considered.

In the first place, it is hardly necessary to argue that economic security, in whatever sense, is important to the worker. Common experience alone supports this. The prospect of losing one's job can hardly be entertaining, the event is likely to be depressing and the experience of continued difficulty in finding any other job may induce in the person the blackest doubts whether he has any place at all in his society. If common experience is to be regarded as scientifically disreputable, the evidence of Bakke (1933), the Pilgrim Trust report (1938) and Rundquist and Sletto (1936) may be cited to show that the fact of unemployment, and especially of continued unemployment, is depressing, if not destructive. This evidence is supported indirectly by findings in the present investigation. In reply to the question: 'What is the worst thing that could happen to you?', 20 per cent of the workers who were asked this (factories A to E only) specified unemployment, which is by far the most frequent single category. Further, it was found that among the older men who had been exposed to the possibility of unemployment (age 35 and over, or 66 of 127), those who had been unemployed specified unemployment as the worst thing that could happen to them significantly more often than did those who had not been unemployed.

The consequences in behaviour at work of the worker's judgement of the security or insecurity of his job would probably depend on the circumstances and it may be that there would be no considerable consequences unless it is obvious to the worker that his job is insecure; in which case covert behaviour might be the chief point of interest, as argued above. In any case, it is necessary to consider why the worker's judgement of job security was not investigated, and this requires a definition of the possible meanings of economic security. Firstly, the worker's job may be secure because it is naturally permanent, or insecure because it is naturally temporary. The job may be one that is bound to elapse in time (because the firm has contracted to complete a certain task, but has no definite commitments beyond that), or one that the worker can expect will elapse in time (because his employers are engaged in seasonal work, such as building or fruit canning), or one that may or may not be required from day to day (such as casual labour in the docks and elsewhere). It may well be that the insecurity of jobs of any of these three kinds has important consequences in behaviour at work. As the investigation described did not include any persons in temporary or casual jobs, no more can be said about this. A second possibility, however, is obvious: that, for the person whose job is in its nature permanent, security may depend on the success or failure of the firm or industry in which he is engaged. His job is secure if his particular firm is likely to keep ahead and beat its rivals (if any); or if the particular industry in which his firm is engaged is likely to remain stable in any circumstances; or, lastly, if the whole economy of his society is likely to remain stable. Conversely, the worker's job is insecure if his particular firm is inefficient and therefore likely to go under, or if the particular industry in which his firm is engaged is exposed, marginal, luxurious or non-essential and therefore likely to be badly hit by a depression; or if he works in a society with a generally unstable economy. To include these three cases under the one general head of economic security or insecurity may be difficult; but, if this is done, one has a third and most general meaning of security which could certainly have been investigated in principle.

In practice, the investigation of economic security in this general sense, or of any of its three possible specific components, would have been difficult. To have invited direct, informed and realistic judgements of the particular firm, the particular industry or the whole social economy would probably have been useless because judgements of this order are beyond the worker's range of information or verbal sophistication. It is not suggested that the persons interviewed were uninterested in their general economic circumstances or incapable of appreciating them. But, in the circumstances of the investigation, detailed judgements could hardly have been expected. The formulation of judgements of general future prospects would have been more difficult, in any case, than the formulation of judgements of the present work situation. The more important point is that the investigation was carried out at a time of full (or even over-full) employment, so that there was no practical possibility of comparing this situation, as judged by the workers, with other situations. It is altogether likely that economic security, as judged by the worker, is an important ground for action in some circumstances; in other circumstances (full employment) it is not only incapable of assessment, it is probably unimportant.

It may be concluded that the opinions analysed in this chapter are, for the most part, considered judgements of the work situation and not merely opinions that would be just sufficient to meet the ostensible requirements of the interview. It follows that these independent judgements are likely to have important consequences in the person's behaviour at work, though these cannot be considered in detail for lack of evidence. Further, the judgements analysed give a fairly complete account of the work situation as it was at the time, and in the circumstances, of the investigation. It follows that the analysis of the effective causes of these judgements, in their order of importance, is certainly worth undertaking. It follows, equally, that no analysis can elucidate the causes of final evaluative judgements of work situations other than those which were studied. As there is no evidence from economically insecure situations, findings must be limited to permanent and economically stable situations. As there is no information from workers who had not yet settled into their jobs, findings must be limited to the settled and, on the whole, older workers. Subject to these restrictions, the analysis reported in the following chapters was concerned with the four distinct judgements of the work situation that were discussed above. It would have been easy to combine these into a single 'morale scale', using the rank weights 4 to 1 for task to mates, in order, but this would have only made the analysis more difficult. The scale so derived would be hard to handle by the methods of parametric statistics, as it would inevitably be multimodal (as can be seen from the appropriate reductions of Table 11). A more important consideration is that a single morale scale would have been difficult to interpret, if not meaningless. The finding that the four judgements of the work situation are independent is as important as the finding that they are probably not equally effective in promoting action at work. The analysis set out in the following chapters therefore ignores the merely statistical possibility of defining one overall judgement of the work situation and proceeds directly to the elucidation of the four distinct judgements.

CHAPTER IV

THE WORKER'S TASK

THE factory worker's judgement of the task he is paid to do presumably depends on the nature of the task. The worker says the task is clean, cool, easy, interesting, or the reverse. Any of these judgements may concur, or not concur, with the judgements of a detached observer. There is the possibility that tasks which have various desirable characteristics are those which are liked, and the converse. This possibility can be tested if the tasks in question are classified by the observer, acting as an independent assessor.

Obvious as this proposal may seem, a little consideration of what it involves is necessary. The observer may not be well placed to make his own classification of tasks because, as noted previously, he may not have been able to conduct systematic and prolonged observations. On the other hand, such observations may have been made by somebody. At the least, there must be some sort of official statement of the duties of any task. If it has been prepared by an industrial expert as a legal document (i.e. as the basis for a wage determination) or for use in a comprehensive training scheme for operators, it will probably be a very detailed statement. Besides this, there may be job descriptions drawn up for more general purposes, such as the selection or allocation of workers, or vocational guidance. These, if they are more likely to refer to groups of similar jobs than to single tasks, will also probably be very detailed. Both kinds of job description will have been made by persons with much better access to the evidence than had the psychologist who fitted in some unsystematic observation of tasks between spells of interviewing.

The more obvious procedure, then, would have been to use official job descriptions where they were available: for all the tradesmen's jobs and for about half of the others. However, these descriptions would not have served the purpose of the investigation, for two reasons. The first and more important is that neither would have been sufficiently specific, or specific on a relevant basis. The job description produced for official use in the factory may indeed be highly specific; it may distinguish exactly, for example, between two machine operators whose work-cycles are identical except for one minor element that is due to differences in machine design. None the less, it may omit elements, especially elements of social relations, which may be relevant to the worker's satisfaction with the job. The worker's degree of contact with his immediate boss and with his workmates (in the course of work) were both considered in the investigation as both are possibly relevant. These are seldom specified exactly, if they are mentioned at all, in even the most detailed job description and they hardly could be specified exactly in a broader description of a range of jobs. Further, some other information may be hard to interpret. Detailed job descriptions often include a full list of physical characteristics, such as heat, cold, dust, noise and so on. Usually, these are simply noted as present or absent. Lacking absolute measures of these characteristics, the job analyst may rate their incidence against a hypothetical idea of 'average' or 'excessive' degrees. But this is not usually stated and, if it were, would not be relevant. For the purpose of the investigation, what must be known is, at the least, which were the cleaner and which the dirtier tasks *in the particular range studied.* This could be known, from official sources, only if it were certain that all of these had been drawn up, in all the factories, on exactly the same basis; which was obviously not so. The same difficulty, of course, would apply even more forcibly to job descriptions of the broader kind.

Secondly, there is the difficulty that the official job description may be too general or official to be useful. The statement of duties for the maintenance engineer, for example, may make no note of exceptionally hot conditions of work because these do not generally apply. But, if one group of maintenance engineers is stationed permanently in a particularly hot corner of the factory, the distinction may be important. Or again, the official statement may describe the task of a particular kind of press operator as clean in all respects. But if the men concerned work next to a dusty section and the prevailing draught blows the dust straight across their presses, a possibly relevant feature may be lost because officially it should not, and therefore does not, occur.

The conclusion is simple: any scheme of job description, analysis or classification is as good as any other so long as it serves its purpose. No criticism is implied of either kind of official job description, for both do serve their different purposes. The employer must know what the worker has to do, and has to be trained to do, and for this purpose the highly specific job description which is usually concerned only with direct task performance is quite appropriate. The firm's employment officer or the official of an employment agency or the vocational psychologist must know, in much more general terms, what the worker is supposed to do; and, for their purposes, the broad description of groups of tasks is quite appropriate. However, the purpose of the investigation was to classify and compare tasks with regard to those elements that might be related to the worker's final evaluation of the task. For the reasons stated, it could not be assumed that official job descriptions (where they were available) would serve this purpose. If one sets aside the fascinating but remote possibility of constructing a general taxonomy or morphology of tasks, one is left with the obvious practical solution: to make a classification suited to the particular case and particular purpose.

This classification must include any purely local features of the task which are observed, even though they may not be officially supposed to be present. It follows that findings will be limited to the particular range of evidence about the tasks as they are actually done. Findings will not necessarily apply to the 'same' task, as it might be regarded in conventional job analysis, in another factory or even in the one factory. But this is no drawback. On the contrary, it is important to know what elements of the task, and therefore what differences between tasks, are related to satisfaction. If it appears that some small and local or casual differences may matter, this is a relevant finding. Equally, the classification must include all the elements that may be suggested by the findings of chapter 3, and not only the obvious physical characteristics of the task. Here, again, the possibility is important that some relevant elements might be of a kind that is seldom noticed in the conventional job analysis. Lastly, the classification must depend on the relatively short and unsystematic observations of the investigators. The point has been somewhat laboured that no other basis would be possible, as it might appear that this basis, if suitable in principle, would be insufficient in practice. In fact, no difficulty was found. All that was necessary was to arrange the tasks observed in a simple dichotomy for each element: those that had more and those that had less of the particular element. For this purpose, the evidence of unsystematic observation was found to be quite sufficient.

TASKS AND SATISFACTION

The procedure was in two stages. In the first, the tasks of the men interviewed were simply listed under their titles or, in some cases, classified together because the operations appeared to be closely similar even though the titles differed. This gave a long list of over 60 different tasks. Then each task was assessed for the presence or absence of each of nine elements, or for the degree of occurrence of each element; the criterion depended on the particular element, some of which were clearly either there or not and some of which ranged by degrees, though not by clearly equal degrees. In the third stage, the long list of tasks was reduced to a final short list of 29 (as shown in Table 13) by the combination of tasks that were identical with regard to the nine elements considered. This list, of course, is purely an *ad hoc* classification, suited to the purpose of the investigation. It retains many distinctions that depend on the nature of the operations but suppresses other distinctions. For example, instrument fitters and pattern-makers (row 8), are classified together not because their task operations are similar but because the nine elements considered are identical for these tasks. For the converse reason, some distinctions are retained that are not given in the nature of the operations. Lastly, it should be noted that the very brief titles of the 29 groups of tasks are not meant to be self-explanatory. Besides that some detail had to be suppressed in order to reduce the possibility of the factory's being identified, only short titles can be given in a table. Elucidation, however, is best undertaken in relation to the findings for the nine elements, and these are best considered in four groups.

(i) Four simple physical characteristics of the task and its environment may be taken first: clean or dirty, light or laborious, quiet or noisy, at normal temperatures or not. The 189 men who were interviewed in factories A to E and G were classified according to the presence or absence of these four characteristics in their tasks (as shown in Table 29 for the first of the four only). These four dichotomies were then tested against the dichotomy of satisfied with the task/not satisfied or dissatisfied. A highly significant positive association was found for the first only, that the task is clean or dirty; the other three possible relations were insignificant. This finding may be analysed in detail with regard to the definitions of the four characteristics. A dirty task was defined as one in which the worker necessarily gets grease, dust, mud or other dirt on his clothes; or on his hands; or in which his clothes, body and lungs are exposed to dust, chemical fumes or vapours; or any combination of these three points. A clean task was defined as one to which none of these points applies. Some of the distinctions shown in Table 13 are obvious, but others require brief explanation, either to elucidate obscurities in the short titles or to draw attention to specific local features of the task.

The distinction between tasks 17 and 18, for example, is not self-evident. There were machinists in control of almost entirely automatic machines (task 17) who had little to do but press buttons and watch the machine work. There were others (task 18) whose extensive hand-feeding of their machines necessarily made their hands and overalls dirty. This did not apply, however, to task 20 (in the same factory as 18 and 19), as in this task the worker made the complete, individually ordered, article from start to finish and in consequence handled tools more than he handled sheet metal. Tasks 21 and 22 are distinguished from task 23 by the fact that in the former two there is very little, if any, handling of raw material (in task 22, this is done, if required, by the cook's labourer); while in task 23 the worker has to work the raw material by hand. Lastly, a purely local point may be noted. In task 6, as it

TABLE 13

Satisfactory elements of tasks

	The task is		
	Clean	Complete	Independent
1. Engineer: production, makes machine parts			
2. instals new and repaired machines			
3. repairs machine parts, in workshop			
4. repairs machine parts, on one location		√	√
5. repairs machine parts, anywhere in factory			
6. Machine setter, automatic lathes			
7. Toolmaker	√	√	√
8. Instrument fitter, pattern maker	√	√	√
9. Welder, repairs machines	√	√	
10. Electrician, maintenance and installation	√		
11. Painter	√	√	
12. Carpenter, construction and repairs	√		
13. Chemical operator: reads gauges, operates valves			
14. mangles products			
15. does several processes by hand		√	
16. mixes powders			
17. Machinist, sheet metal: small articles	√	√	
18. large articles			
19. Spot welder, riveter			
20. Bench hand: makes sheet metal products by hand	√	√	√
21. Food finisher	√	√	√
22. Food cook, various	√	√	√
23. Food machinist			
24. Food machine attendants, cook's labourers			
25. Storeman	√	√	√
26. Fork-lift truck driver	√		
27. Dispatcher, packer, bagger, loader	√		√
28. Yardman, cleaner			√
29. Tradesman's labourer, general labourer			

was observed, the setter spent a good deal of time not only supervising the lathe operators but also making minor running adjustments which he thought were too difficult for the operator. In the nature of the task, however, this is not inevitable.

The three physical task characteristics that were not related to satisfaction with the task can now be considered. The first of these, whether the task was light or laborious, is probably unrelated only in the circumstances of the investigation. For this characteristic, the only distinction that could be made reliably was between the very few tasks that obviously required gross physical effort ('heavy bullocking work', to quote a worker, at a probable total output of one horse-power) several times each day, and all the other tasks. A comparison over a wider range of tasks, including more that required gross effort, might well have given a different conclusion. So might a finer set of comparisons in the range of the investigation, but this would have required physiological measurements of the sort reported by Christensen (1953), and not merely observation. Much the same goes for another characteristic that was found not to be related to satisfaction with the task, the temperature and humidity of the environment. At the time of the investigation (early autumn to early spring), there were very few tasks that had to be carried out under conditions that were markedly hot and humid, and none at all that appeared even to approach extreme conditions. Here again, accurate measurement in the limited range taken, or observation over a wider range, might well have given a different conclusion. The case with the last unrelated characteristic, the level of noise of the task, is different. In three of the six factories, sections were easily defined in which conversation at less than a steady bellow was impossible, at least for persons not adapted to the noise. It was the classification of the workers that was uncertain. There were maintenance tradesmen who happened to have been working for some time in noisy areas, but who were not permanently tied to these areas, and other tradesmen who had never entered them. Some mobile labourers, again, entered noisy areas frequently, others occasionally, others never. Thus several classifications of the workers were possible, and were tried. The fact that no relation to task satisfaction was found with any of these may be due only to the difficulties of making an accurate classification, taking into account frequency and regularity of exposure as well as intensity, merely from short-term observations.

(ii) The next consideration is two less obvious physical characteristics of the task. Firstly, the worker may do a relatively complete task or a fragment only of a complex serial task. He may make a complete product, through a complex series of operations from start to finish; he may carry out the last operations which transform a raw product into one which is visibly ready for use or sale; or he may completely re-make a machine and get it back to work. If he does any of these, his task is relatively complete; if none, it is relatively fragmentary. The distinctions may be taken in detail. There are cooks (task 22) who make a product from the raw ingredients through to the stage where it is ready to be packed; and these may be contrasted with the machinists (task 23) who carry out only one of three to six major stages of cookery, or with the labourers (task 24) who merely assist the machinist or cook. There are the bench hands (task 20) who make an individually ordered article from start to finish, contrasted with the sheet metal workers (tasks 18 and 19) who do one operation only. Among the chemical operators, there is a direct analogue to the cooks. There are men (task 15) who, in the course of the day, mix and cook chemical ingredients through a long and varied cycle of operations, seeing and sometimes handling the products at intermediate stages as well as at the end. These may be contrasted with operators who carry out one process only (tasks 14 and 16) or who, if they carry out more than one process, do so by the remote control of unseen and usually unknown materials (task 13). The toolmaker (task 7) makes a wide variety of sets of tools in small numbers, while the lathe-setter (task 6) merely sets and then adjusts one kind of tool at a time; a contrast, it should be noted, which depends heavily on local practices and conditions. Lastly, there are the instrument fitter and the pattern maker (task 8), whose work is obviously complete (though in most circumstances the instrument fitter would spend most of his time re-making rather than making). So much, then, for the makers and non-makers.

The painters (task 11) finish the task, as they come last on all new or repaired constructions; and the installation engineers, electricians and carpenters (tasks 2, 10 and 12) succeed each other in merely preparing the way for the construction to be finished. The men who do task 21 must be included here, as the actual task (which unfortunately cannot be detailed) is quite similar to that of the painter. The same goes for some sheet metal machinists (task 17), who operate the machines which give the product its final shape and articulation. Lastly, there are the storemen (task 25), who assemble varied packs of products, following a complex list of individual orders (and who could therefore perhaps be regarded as makers); and they may be contrasted with the men who pack one kind of thing only or who merely take products from place to place (tasks 26 and 27). These are the finishers and the preparers.

Some engineers (task 4 only) re-make machines from start to finish by dismantling the machine, making replacement parts (small and simple ones only) in their adjacent workshop and re-assembling the machine, ready to go. These may be contrasted with the engineers (task 5) who are limited to replacing on the spot the very simplest parts only, that they bring with them, and most of whose work therefore has to be referred back to the remote main workshop. This distinction, of course, depends solely on certain local practices and much the same is true for the assessment of the welders (task 9) as re-makers; it so happens that their work usually goes through to the stage where the machine is again fit for use. The assessment of the other engineers (tasks 1 and 3) is clearer, as both make only single parts of machines, whether for new machines or as replacements. The distinction here is between the re-makers and the processers.

The men not mentioned so far (tasks 28 and 29) are obviously non-makers of the lowest order, though in some cases they could be classified with the preparers or processers.

This elaborate classification of the tasks of 189 men was reduced to the simple dichotomy between complete and fragmentary tasks, and a highly significant positive association was found between this and satisfaction with the task.

The second of the less obvious physical characteristics of the task is mobility; the worker may move around the factory in the course of his task or he may stay in one place. This is easily assessed, but is not shown in Table 13, as it was found to have no relation to satisfaction with the task. There is no apparent reason why it should, except in so far as it implies independence; but this, which is discussed below, depends on other factors besides mobility. The analysis of the physical characteristics of the task is therefore concluded with this sixth and last possible factor.

The worker's final judgement of his task, however, does not necessarily depend only on the physical characteristics of the task, whether on the two most obvious or, in a more detailed analysis, on others besides these. To assume that only factors of this kind can be relevant is to concede too much to the conventional job analysis in terms of physical operations only. It is at least conceivable that the social relations which are a necessary part of the task may be relevant. So, equally, may be pay, as the necessary counterpart of performance in the task. These two possibilities must be considered in turn.

(iii) The simpler possibility may be considered first, that there is some relation between the worker's pay and his judgement of his task; and, for this purpose, the worker's pay must be defined. There is a prescribed basic or general rate of pay for each of the tasks set out in Table 13 (and different rates for some workers who are there classified together on the grounds noted above). Beyond this legally prescribed rate, there are wide variations. Some workers are paid a regular or occasional production bonus, or are paid a small margin for regular attendance. These extra payments depend on the general policy of the factory, and are therefore not simply related to tasks or task groups; for example, some engineers (tasks 1 to 5) receive both kinds of extra payment, some receive one or the other, and some neither. Then there are local variations, such as the weekend stand-by pay of some, but not all, maintenance tradesmen, in one factory only. Lastly, there is the occasional variation of payment for overtime, which was open to many workers at the time of the investigation, but which depended on industrial market considerations, and therefore cannot be simply related to tasks or groups of tasks. In the analysis of the worker's judgement of his pay (chapter V), both the legally prescribed rate and the several variations above this must be taken into account. For the present purpose, the whole complex could be taken into account on an individual basis: the relation of pay, however it is inclusively defined (see chapter V), to satisfaction with the task could be analysed regardless of the classification of tasks set out in Table 13. This would be pointless. The question proposed in this chapter is the relation between the task and the worker's judgement of the task. Tasks may be defined in many ways, and the definition from which the classification of Table 13 derives is only one possibility. However, it does depend largely on the necessary operations and environmental conditions of the task. If it were to take into account the several variations of the prescribed rate of pay for the task, the classification would come close to having one category per person. The practical solution is to confine the consideration of pay, in relation to task satisfaction, to the bare necessary counterpart, the officially prescribed rate for each task (or group of tasks). Over the range of 29 tasks or groups of tasks there were considerable differences (at the extreme, about 3 to 2) in the prescribed rates. It was found, however, that there was no relation between these rates and the worker's judgement of the task. As there is no relation between the workers' judgement of his task and his pay, this is not surprising.

(iv) There remain to be considered the last two of the nine elements of the task, the worker's relations with his mates and with his boss or bosses. As with pay, it would be irrelevant to consider these relations on an individual basis (as in chapter VI); but the relations which are necessary to the task are relevant. These are easily defined. The number of persons with whom the worker carries out his task depends simply on the kind of task; and so does the extent to which he is supervised. There are some tasks which are carried out alone, others which require two or more persons, and others in which the number of persons working together varies from time to time. Therefore several classifications of this element in the task are possible, but it was found that there is no significant relation between the number of the worker's mates, however these are classified, and his satisfaction with his task. This finding may be due to the fact that, except under extreme conditions, the worker's possible contacts with other workers may not be limited to those that are necessary to the job; there are often other workers adjacent or passing by. Besides this, the day is divided by the dinner break and at least one tea break, in which the worker can talk to anyone at hand, whatever his task relations may be. The worker may or may not want the company of his mates but, in any ordinary circumstances, he can have it if he does want it.

The second of the social characteristics of the task is the degree of contact that the worker has with his immediate boss, in the nature of the task, regardless of individual differences in supervisory methods. The tasks which are relatively independent are shown in the last column of Table 13. In some cases the worker is independent simply because his task takes him away from his boss. The engineers who do task 4 were placed in pairs on a number of widely scattered locations and it was physically impossible for their foreman to do more than visit them once a day, unless he was specially called for. This may be contrasted with the conditions of the tradesmen (tasks 2, 5, 10, 11 and 12) who move around in the course of their work. For the most part, these men work for limited periods (a few hours up to two or three weeks) on one location; and, for the most part, they work in a small number of large gangs. In consequence (as determined by observation), they are visited by their foremen several times a day at least; the limiting case observed is that the foreman spends half a day with each of his three or four gangs in turn. The mobile labourers present a simpler case. Some (task 27) move around widely in the course of their work, others (task 28) may work on one location for some time, but their various locations are very widely spread, over the whole of the factory's buildings and grounds. In both cases, close supervision is physically impossible. This does not hold, however, for the fork-lift truck drivers (task 26), in spite of their being mobile. The reason for this is purely local: the drivers interviewed had a short and clearly visible run, up and down the central gangway of one factory, passing their foreman's office on each run.

In other cases, the worker is independent because the adequacy of his performance can be directly assessed only at long intervals, or can be assessed only by its relatively long-term results, unless sample checks and elaborate progress inspections are carried out. In tasks 20 and 21, for example, the finished product can be assessed by the foreman only after a whole day's work (task 21) or up to a week's work (task 20, in the most complicated cases). Even then, unless the product is grossly and obviously defective, a final assessment must wait on the customer's report, if any. In tasks 7 and 8, the finished product may take from a few hours to several weeks, but its adequacy in all cases can be effectively checked only by its being put to use. In tasks 22 and 25, the product is finished within minutes (25) or within hours (22); but, except for laborious sample checks, the only possible assessment of its adequacy is the eventual customer's report. These are all complete tasks, as defined above, whether making or finishing. Not all complete tasks, however, can be assessed as independent. In one case (task 9) the product (in this case, a machine repaired by welding) is usually finished in a short time, such as half a day, so that the adequacy of the worker's performance can be tested within the one day. In two cases (tasks 11 and 17) progress is visible at all times. In one (task 15), it is checked elaborately at short intervals because the cost of any one mistake in the complex cycle (spoiling a whole batch of chemicals) would be relatively high. The same holds, of course, for task 13, but this is an incomplete task anyway. Lastly, then, all the incomplete tasks are dependent on the immediate boss (usually the foreman) in the sense that their progress can be assessed by simple visual inspection at short intervals.

The account of the task characteristics that are associated with the worker's final judgement of his task is now complete. It is found that workers whose tasks are clean, complete or independent (as defined above) are significantly more satisfied with their tasks (however satisfaction is defined in detail), than are other workers. As this conclusion obviously depends in many details on particular local features of the investigation, one must add, at the least, that other simple physical characteristics of the task might well be found relevant over a wider range of tasks. However, if the three findings are limited, they suggest three further questions. Beyond the bare fact of statistical significance, the practical effectiveness of these three characteristics may need consideration. Besides this, their effectiveness in combination must be considered: and this may depend on the relations between them.

The answer to the first question is that the associations observed between task characteristics and satisfaction with the task are highly significant ($P<0{\cdot}01$) but of little practical effect. The coefficient of association (ϕ) is 0·25 for the clean/dirty division of tasks, 0·22 for the complete/incomplete division and 0·20 for the independent/dependent division. The possibility remains that the three characteristics are of more practical significance in combination. This depends partly on the answer to the third question, what are the associations among these three characteristics. They are relatively high. The coefficients of association are: clean and complete, 0·59; clean and independent, 0·47; complete and independent, 0·52. In view of these findings it must be expected that the effectiveness of the combined characteristics (the second question) will not be much greater than that of the single most effective characteristic. Table 14 shows in detail the relation of the worker's judgement of his task to the eight combinations of the three characteristics of the task; and, for convenience, includes a note of the tasks, as set out in Table 13, which are included in each of the combinations. The obvious division of this table is between rows 7 and 8: the workers whose tasks have at least one of the three characteristics tend to be satisfied with their tasks, while the workers whose tasks have none of the three characteristics tend not to be satisfied. The association is highly significant ($P < 0{\cdot}001$) but is of limited practical efficiency, as is shown by the coefficiency of association (ϕ), which is 0·28. A more concrete statement is that the generalization proposed holds for 121 of the

TABLE 14

Characteristics and judgements of the task

The task is	The worker's judgement is			The tasks included are
	Satisfied	Not	Total	
1. Clean, complete, independent	32	5	37	7, 8, 20, 21, 22, 25
2. Clean and complete	8	2	10	9, 11, 17
3. Clean and independent	16	5	21	27
4. Complete and independent	5	2	7	4
5. Clean only	12	3	15	10, 12, 26
6. Complete only	5	1	6	15
7. Independent only	3	1	4	28
8. None of these three	49	40	89	1, 2, 3, 5, 6, 13, 14, 16, 18, 19, 23, 24, 29
Total	130	59	189	

189 persons, or 64 per cent. It may be added that there are no clearly significant differences in this association between the combined task characteristics and the worker's judgement of the task by nationality or occupational grade. The final generalization, however, holds more strongly for the older workers (age 40 and over) and for the workers with long service (2 years and over) than it does for workers who are younger or who have shorter service.

The general finding on tasks is obviously limited in various respects by its derivation from the particular range of tasks observed, so that it cannot be extended without weighty reservations to include tasks outside this range. Besides this, it is a finding of no more than a statistically significant tendency, of no great practical effect. Besides this, again, its range as a general finding is apparently limited, in that it is clearly true for the older and more experienced workers, but not so clearly true for the others.

None the less, it is an intelligible finding, not to say a simple and obvious conclusion. Subject to the reservations noted, the worker apparently judges his task in much the same way as his other, relatively free or unprescribed activities. The task is satisfactory if it is clean, or complete, or independent, or any combination of these. But these are the characteristics of the activities in which all persons engage outside the sphere of work. The person may indeed, in one sense, work very hard at being a husband or father, or at making things, or at repairing things, or at expressing himself, or simply at playing. In whatever case, he does not take on dirty work unless there is an over-riding reason for doing so. Nor does he take on purely repetitious work; in the nature of the case, whatever he takes on he must do himself from start to finish. Lastly, he is obviously independent because he works for himself alone. The worker's task, then, is satisfactory in so far as it approximates to some or all of the conditions of his other activities and unsatisfactory if it lacks all of these conditions.

This rather general and speculative conclusion is not easily compared with other findings, even if it is reduced to its components. The findings on the physical characteristics of the task agree broadly with those of Walker and Marriott (1951), though the latter include characteristics that may be important in some circumstances but which were beyond the range of the investigation. That a satisfactory task is a complete task is an opinion which has been put forward, with some support from laboratory psychology, by many industrial psychologists; but a comparison which depends on detailed evidence is not possible. Walker and Marriott, again (1951), note that repetitious tasks are less satisfying than are other tasks. The difference found, however, is not directly comparable because there are incomplete tasks (such as tasks 1, 2, 3, 5 and others in Table 13) which could hardly be described as repetitious. The importance of the task's being independent seems not to have been noticed previously. As this depends solely on the nature of the task, it cannot be related to the various findings about the influence of personal differences in supervision on the worker's satisfaction with his task.

CONCLUSION

It appears that the factory worker makes a set of reasonable and intelligible judgements of his task on its three main characteristics. The last point to consider is on what comparative basis these judgements are made. That they are not made on an absolute basis is evident. No worker, and probably no independent assessor, could exactly measure tasks for their cleanness, completeness or independence on an absolute scale ranging from zero. What seems likely is that judgements are referred to, or compared with, whatever would be open to the worker as he is. This includes all tasks that he could now undertake: within his particular trade if he is a tradesman; or within some range, whether limited or very wide, of semi-skilled tasks if he is a process worker; or within the range of labouring tasks (and possibly semi-skilled tasks) if he is a labourer. A minor qualification is that judgements may be made on the basis of what would have been open to the worker before he made the definitive choice of his particular trade. This would apply only to persons who are now tradesmen, and whose previous decision was therefore for this or that particular trade. For other workers, the first major decision both blocks off completely a wide range of possible tasks (the trades) and opens up a different but also wide range of semi-skilled or labouring tasks.

It seems likely, then, that no worker judges his present task in relation to the whole range of factory tasks, many of which must be unknown to him. His judgements are probably related to his actual or hypothetical possibilities of movement from one task to another. Thus the task is cleaner than some other tasks he might do, within his trade; or cleaner than tasks in some other trades he might have entered; or cleaner than the tasks of the process workers and labourers with whom he works now or with whom he may have worked previously. The same would hold for the other aspects of the task: completeness and independence, equally, are presumably judged in relation to both present possibilities and past hypothetical possibilities, where those matter. The possibility of downward movement among the set of factory tasks may be ignored as a ground of judgement. In a sense, it is open and occasionally occurs. For example, the tradesman may decide to abandon his trade meantime because he can make a better living as a process worker or even as a labourer. In general, however, it appears that the worker who takes on a task at a lower level than he must does so only if he must; the possibility of the tradesman's becoming a process worker or of a process worker's becoming a labourer is not open in so far as only the dire pressure of circumstances would suffice to make the person accept a dangerous occupational declassification and (usually) an immediate drop in pay.

The last point that must be made is that the worker's judgement of his present task, if it depends on his possibilities, must depend also on his experience, or his previous tasks and his judgement of these. This is at least suggested by the findings that the older workers and the workers with longer service are those who make the most intelligible judgements. It may be concluded, then, that the worker assesses the three main characteristics of his present task in relation to his present or hypothetical possibilities as a worker and to his previous experience of tasks. As a reconstruction of the worker's covert behaviour, this is at least plausible. The fact remains that this covert behaviour, or set of judgements, leaves much of the worker's final evaluation of his task unexplained. The worker's judgement of his task depends significantly and intelligibly, but does not depend only, on the characteristics of the task as assessed by the independent observer. The other possible antecedents or determinants of this judgement remain to be considered in later chapters.

CHAPTER V

THE WORKER'S PAY

THE worker's judgement of his pay depends, of course, on his pay; but not necessarily in a simple and obvious sense. Pay is not a thing or a process with visible characteristics; it is a means of exchanging services for goods or other services. It is both a reward for doing something and an instrument for doing something else. At least four factors in the worker's judgement may therefore have to be considered.

His judgement may depend, in the simplest case, on the rate of pay, whether on the official base rate or on his actual take-home pay, which are alternative specifications of the gross size of the reward. Or the judgement may depend on what extra payments the worker gets; that is, on the detailed components of his total pay, or on the relation between the base rate and the remainder. Thirdly, the judgement may depend on what the reward is for, on the kinds of task for which the worker is paid. Lastly, the uses of the instrument, or the worker's financial responsibilities, may be relevant. Doubtless other possibilities could be suggested. These are the factors that seemed most likely to be important, and which were therefore investigated.

With pay, the difficulties of definitions should not be nearly so great as with the task, as definition does not depend at all on the observer, but only on the collation of interview reports together with official records and statements. None the less, there were considerable difficulties. Detailed official information was available only for factories A to E (that is, for 127 men), but among these five factories alone the pay systems were both different and complicated. It was found necessary to distinguish ten elements, as set out below.

1. The award rate, or legally defined rate, for the particular task.

2. The fixed margin, if any, paid above the award rate throughout the factory. Where there was such a margin, the amount depended on the official job classification but was the same for all persons in any one group, and was therefore part of the particular factory's base rate of pay for the particular job.

3. Another fixed margin was staff pay, which was paid in one factory only to certain men who were working as tradesmen rather than as supervisors. This is distinguished from element 2 because it was regarded as an extra payment that could be withdrawn and not as part of the locally prescribed base rate.

4. Standby money was paid to some tradesmen in one factory, who were required to stay at home on specified weekends so that they would be available for emergency repair calls.

5. In two factories, attendance money was paid; a small weekly margin for regular attendance.

6. Extra pay for shift work occurred in one factory. This margin, of course, is officially defined in the same way as element 1, the award rate. It is specified separately because it was possible for a man who did not like shift work to go on to day work instead, and so to do exactly the same task at a lower rate of pay. From the worker's point of view, it was to some extent an optional extra.

7. In one factory, there was a substantial bonus (up to 18 per cent of the worker's base pay), that depended on the firm's profits for the year. That it would be paid to everyone in the factory shortly after the end of the year was certain, being the declared policy of the company. The amount was not known till it was paid.

8. In another factory, a production bonus was paid to one section only of process workers. It was paid weekly, and was expected by the men concerned; but the amount was unpredictable, as it depended fully as much on preceding processes as on the efforts of the men in the section.

9. In two factories, a production bonus was paid to some workers, either yearly or half-yearly. In both, no definite rules were laid down to determine either which workers should get a bonus or how much it should be. In consequence, the workers concerned did not expect a fixed amount or proportion of their wages (cf. element 7), or any bonus at all. At the most, they hoped to get something extra in due course; and, at the least, they had merely heard that extra payments had sometimes been made, on this basis, to unspecified groups of workers.

10. Payments for overtime work occurred in all five factories, according to the official rules; usually, time and a half for the first four hours and double time thereafter (including public holidays).

This list of ten elements in the factory's pay systems may be reduced to four main combinations. Elements 1 and 2 together give the local base rate of pay for any specified task. Elements 3 to 6 are all extra payments that are fixed in amount and known to the worker and that are paid regularly and frequently (weekly, except for element 4, which depends on a fairly long cycle of rotating duties). Elements 7 to 9 all vary unpredictably in amount and, for the most part, are paid at long intervals. Element 10 only depends directly and predictably on the worker's own efforts.

The obvious source of information for the definition of all ten elements was the pay records of the factory. It was found, however, that these were of limited use, for two reasons. In two of the five factories, the pay section would have been hard put to it to produce a record, for each of the workers interviewed, of his total take-home pay over a substantial period. In all five factories, there was no possibility of analysing this total (even where it was readily available) into its various elements without undertaking a separate project on this point only. The pay records of all factories, of course, were kept perfectly for taxation and other official purposes; but these purposes, unfortunately, are little related to the interests of the research worker. The local base rate of pay (element 2) and the fixed extra payments (elements 3 to 6) could in all cases be specified in general from the factory records. It was not possible, however, to specify in all cases exactly which persons were entitled to, or had been receiving which extra payments (element 4 in particular, and also element 6, occasionally). It followed that in some cases in which the official records could be analysed into elements, the incidence of these elements had still to be checked with the interview records.

These latter were the obvious alternative sources of information about pay, but were found to be less reliable on the whole than the official records. The worker's statement of his total take-home pay (excluding only elements 7 and 9, which were not regarded as pay) was precise in all cases; but in many cases it was subject to the qualification that pay varies with overtime (element 10) and with other elements. Besides this, there were considerable individual variations in what was reported as basic pay and what was reported as extra payments. Some workers carefully specified element 1 only, the legally prescribed minimum rate, as their pay, while others included all the predictable extra payments (elements 2, 3, 4, 5, and 8); and so in between these extremes. In all cases, some workers stated exactly the amounts of the relatively fixed extra payments (elements 3 to 6) and of their personal overtime earnings (element 10), while many others had little idea of the details that made up their actual take-home pay. In this case too a considerable research project would have been required to determine the details; and in this case particularly, more detailed investigation would have been impossible within the general terms of the project. It was therefore concluded that both sources of information, factory records and statements made in the interview, must be used as they were and checked against each other as far as possible.

RATES, EXTRA PAYMENTS AND THE TASK

The first step in the analysis of the workers' judgements of their pay was to define their rate of pay, or gross amount of weekly pay.

Two definitions were made. From the factory records, the worker's relatively fixed base rate of pay was defined as the sum of elements 1 to 6. Total take-home pay was defined as the sum of all elements except 7 and 9, which were very seldom regarded by the worker as part of his pay. This definition, derived from interview records, was individual, but the definition from factory records applied to groups of workers indiscriminately. The relation between the two definitions could therefore be stated only in terms of association, and was reasonably close ($\phi = 0{\cdot}68$). For the official definition, which was the more reliable, the range in rates of pay from the highest to the lowest paid group was considerable, in the ratio of 1·64 to 1. For the less reliable definition from interview records, the range from the highest to the lowest paid individual was greater. The extreme ratio of 2·65 to 1 depends on one man who was working two shifts as a process worker, in two different tasks in the one factory. Excluding this one man, the ratio from the highest to lowest is still 2·05 to 1. The discrepancy between this and the ratio for pay groups (or task groups) is to be expected, as the estimates for individuals include more information, and especially more variable information, than do the estimates for groups. Even so, the estimates for individuals must be taken to be the less reliable. As a fact of history, the amount of pay that the worker took home last week is indisputable. It was found, however, that a considerable minority of workers did not specify their take-home pay exactly even when it was relatively constant. Doubtless they could have been exact; but there was no time, in the interview, to pursue the nice distinctions among official rates, extra payments, and total effective pay. More important than this was the difficulty that for many workers the effective (or take-home) rate of pay varied considerably from time to time, depending on how much overtime they worked. Here also, it was not possible in the interview to elucidate the full historical details of how much overtime had been worked, or was usually worked. Both the group and the individual estimates are therefore subject to error, but both should be sufficiently accurate for use in analysing the elements of satisfaction with pay.

It was found that neither estimate is associated with the worker's report of satisfaction or dissatisfaction with his pay. This may be because both estimates are unreliable. However, the association between the two suggests that this cannot be the sole explanation. The worker's judgement of his pay seems likely to be little related to the simple amount of pay, on whichever basis this is estimated. It remains possible that the judgement is related to the relative size or kind of extra payments or the kind of task for which the worker is paid.

The second stage of the analysis therefore included only payments above the base rate (elements 1 and 2). Because of the difficulties found in defining both the base rate of pay and the effective rate, it was decided to define payments above the base rate without regard to the amount, solely according to whether they were made or not made. This simplified the treament of the data. For elements 3, 4, 5, 7 and 8 the official records showed clearly to which groups or individuals extra payments of these five kinds had been made, or were usually made; and there were no discrepancies with the interview records. For element 9, the interview records were taken as primary, but were checked against the statements of factory policy where possible; as the latter were unclear on the individual application of production bonuses, they were regarded as secondary sources only. For element 6, only the interview record mattered. If the worker at the time of interview was, as he said, on shift work, then he was paid the prescribed margin for this: that he may have been a day worker previously was taken to be immaterial. Lastly, the interview records were the relevant source for element 10. Whether or not anyone works overtime varies considerably, not only from section to section within any factory but also from time to time for any one section. The rule was therefore adopted that persons were regarded as working overtime under any one of three conditions. That they reported doing so regularly (one or two nights every week) was the simplest condition. Occasional overtime, as variably required by the particular task, was also included; and so, lastly, was regular or occasional overtime worked in the very recent past (up to a fortnight before the interview). The only persons regarded as not working overtime were the few who stated firmly and definitely that they had not done so, and would not do so because they were against it.

The findings from this stage of the analysis are simple. There is a highly significant positive association ($P = 0{\cdot}002$, $\phi = 0{\cdot}300$) between satisfaction with pay (or the worker's positive judgement of his pay) and working overtime. For all the other elements considered, the association is insignificant ($P > 0{\cdot}05$ to $= 0{\cdot}99$). These finding are not remarkable. Payment for overtime is the only general extra payment that occurs in all factories; the other local and particular extra payments are relatively infrequent (element 10, overtime, includes 108 of 127 workers; elements 3 to 9 include 6, 12, 18, 19, 18, 36 and 41 of 127 workers respectively). A more important point is that, within the range of the investigation, overtime pay alone depends on the worker's decision. Apart from the few persons whose tasks never require overtime, the large majority of workers can work overtime if they wish, but (in

1952, at all events) only if they wish. There is no other payment above the base rate which is in the same way open to individual initiative or which is equally profitable; it must be recalled that overtime work is relatively highly paid. A final point is that overtime work yields obvious and visible results that are directly linked to the worker's effort, which is not true of any of the other kinds of payments above the base rate.

None the less, it should be possible to reduce the confusion of payments above the base rate (other than overtime) to a simple order. As these other extra payments depend very much on the policies and practices of the five factories concerned, the obvious reduction is simply to count the number made to each worker, regardless of amount (which is unreliably specified anyway) and of origin in particular factory policy.

When this reduction was made, it was found that the total number of extra payments other than for overtime work was 2, 1 or 0, with frequencies 52, 47 and 28 of 127 workers. This is a reduction to a scalar form of the local and highly variable extra payments. It was found that there is a significant relation ($P = 0{\cdot}031$) between this scale and the worker's judgement of his pay: workers who get 2 extra payments, of whatever kind, tend to be satisfied, and workers who get 1 or 0 tend to be dissatisfied.

A simple conclusion may be drawn from the analyses discussed so far. The worker's judgement of his pay does indeed depend on his pay, but it depends on how his pay is made up and not simply on the amount. It seems likely that the worker accepts the base rate of pay as a given fact which can hardly be changed by his immediate efforts. The official rate for his occupational level in general, or for his specific trade or process, is what it is in the particular factory which employs him, and is neither satisfactory nor unsatisfactory in itself; it is a fact, which may or may not be capable of modification. In the same way, his total effective (or take-home) pay is a fact which is neither satisfactory nor unsatisfactory in itself; what matters is the relation between this and the base rate, or the ways in which the latter has been added to.

Two kinds of addition to the base rate are regarded as satisfactory. Payments for overtime work depends on the individual worker's efforts, but does not require a remarkable effort as much overtime work is paid at twice the ordinary rate. Besides this, the worker can predict exactly what he will make by working overtime. As he may see it, this is the most rewarding part of his weekly labour. The other extra payments are different, in that the worker cannot know in advance how much he will get (in most cases, in the particular range studied) but does know that he will get whatever it is almost regardless of his individual efforts. Among the seven kinds of extra payment that were summed to give the number received, element 3 is almost a guaranteed part of the worker's base pay, as locally defined. Elements 4 and 5 require the worker to observe certain rules, but these are not onerous in nature. Elements 7 to 9 are, in a sense, incentive payments; but none of them, not even element 8, is directly linked to individual effort over short intervals. Element 6, lastly, requires that the worker accepts a tedious monthly routine but not that he exerts himself, on the task, more than any other worker. Finally, the two kinds of extra payment are not related in the limited range of cases studied. In this range, the person may work overtime and receive two extra payments above the local base rate, or either alone, or neither, indifferently. The final conclusion, then, is that pay is judged to be satisfactory if it can be increased by relatively little effort, or by almost no individual effort at all.

The next possibility that must be considered is that the worker's judgement of his pay depends on what he is paid for: that is, on the nature of his task. In this, the third stage of the analysis, there are no difficulties of definition; the classification of tasks as clean, independent or complete (or the converse), as in chapter IV, may be taken as sufficient. The case for considering the relations of these task elements to the worker's judgement of his pay depends on empirical as well as general grounds. The first two elements, cleanness and independence, were found to have a highly significant negative association with the base rate of pay. It appears that cleaner jobs are paid less, which presumably only reflects the official (general or local) margins for dirty (dusty, smelly, greasy, etc.) tasks of any kind. That independent jobs are paid less is probably a reflection of the particular range of cases studied; many of the independent tasks (in Factories A to E, N = 127, but not in factory G, N = 62, which is not under consideration in this chapter) were independent in the simplest sense, that the worker moved around in carrying out his task. That complete tasks were paid the same rates, on the average, as part-tasks again only reflects the features of the cases studied: complete tasks, as studied, did not for the most part require more skill or training than part-tasks. At all events, there is the possibility of analysing factors that are opposed to, or at the least independent of, the factor of amount of pay.

The findings from this, the fourth stage of the analysis, are simple. There is no significant relation between the cleanness or completeness of the task and the worker's judgement of his pay. There is a significant relation (P = 0·05) between independence and the worker's judgement; workers who do independent tasks tend to be satisfied with their pay and others tend to be dissatisfied.

The relation is not of a high order, so that it could well be a statistical artefact. Further analysis suggests that this is unlikely. If the 127 workers in factories A to E are divided into those who did overtime work and those who did not, the relation between independence or dependence in the task and the judgement of pay can be determined for the two sections of the whole group separately. It is found that the two relations do not differ. This is to be expected, on statistical grounds; only 19 of 127 workers do not work overtime, so that a reliable comparison of the two sections is hardly possible. A more important point is that working overtime (or not) and doing an independent task (or not) are not associated at all ($P = 0{\cdot}96$). Thus there is no reason to expect that the relation between independence or dependence of the task and the judgement of pay would differ in these two sections of the whole group. The case is different for the payment of extras other than overtime. A significant negative association was found ($P = 0{\cdot}036$, $\phi = -0{\cdot}204$) between the task-element and the definition of extra payments; that is, persons who do independent tasks tend to have one payment or none above the base rate and persons who do dependent tasks tend to have two payments above the base rate. The relation between independence or dependence of the task and the judgement of pay is significantly closer for the workers who have one extra payment at the most than it is for those who have two. It therefore seems reasonable to conclude that this relation is not an artefact but a true relation which is independent of, and to some extent contrary to, the relations found between pay and the judgement of pay. The independence of the task may therefore be regarded as a substitute, in some circumstances, for extra payments. The three factors in the worker's judgement of his pay are then as shown in Table 15.

The obvious point of division of this table is between rows 4 and 5. The worker is satisfied with his pay if he can work overtime and if he also gets two payments, of whatever kind, above the base rate (rows 1 and 2). He is satisfied with his pay if he can work overtime or if he gets two payments above the base rate, provided that his task is independent, which substitutes for extra payments of either kind (rows 3 and 4). He is not satisfied with his pay if he can work overtime or if he gets two payments above the base rate, but does a dependent task (rows 5 and 6). Lastly, he is not satisfied with his pay if he gets no extra payments, whether for overtime or of other kinds (rows 7 and 8).

This summary statement may be further condensed. Pay, as the worker sees it, is satisfactory if it can be considerably improved by little or no effort or if he is let alone to do his task; it is not satisfactory if there is little scope for improvement or if his task is

TABLE 15

Characteristics and judgements of pay

	The worker reports these characteristics			The worker's judgement is		
	Works overtime	Is paid two extras	Independent task	Satisfied	Not	Total
1.	√	√	√	8	0	8
2.	√	√		33	10	43
3.	√		√	19	3	22
4.		√	√	1	0	1
5.	√			22	13	35
6.		√		0	0	0
7.			√	2	3	5
8.				3	10	13
			Total	88	39	127

closely supervised. This final statement holds regardless of age, length of service, nationality or occupational grade. It cannot be compared directly with the findings of other investigations, because there appear to be no directly comparable investigations. However, the findings of investigations of incentive pay schemes may be considered, as set out by Viteles (1954, ch. II). It appears that workers who are paid directly in proportion to their individual efforts may or may not be satisfied with their job in general, depending on the circumstances. In the case taken, there was no direct payment by results: elements 7 to 9 depended on production, in various defined or undefined ways, but in no case did they depend directly on individual daily or weekly production. In this case, satisfaction with pay (and therefore with the job in general, as distinct from the specific task) seems to depend on the possibility of earning more, or of having a substitute for more pay, with little or no extra effort. It is at least possible that this finding would hold in all cases, so that schemes of incentive payment would be effective or acceptable only in so far as they did offer just this possibility, of earning substantially more for little more effort. It is a commonplace of methods engineering that the chief point of dispute in all incentive pay schemes is not the new and more efficient task routine, but the proposed base rate and the proposed rate of increments above the base. That an acceptable scheme of incentive pay depends on the possibility of earning more for minimal extra effort is a speculation. It is a speculation, however, which is in line with the general findings discussed above; and the converse case, that effective incentive pay would be directly linked to individual effort, cannot be supported at all from these findings. A final point may be made about the kinds of extra payments studied, other than payment for overtime work. These payments depend on the policy of the particular factory, and in some cases it appeared that the factory management regarded the payments made as being of distinctive importance in promoting high morale or job satisfaction. From the findings discussed above, there is no ground for drawing distinctions among kinds of extra payments. It would seem that what matters, from the worker's point of view, is only whether or not extra payments, however they are defined, are made.

PAY AND COMMITMENTS

The analysis of pay as a reward, of what is paid and for what it is paid, concludes with this general finding. It remains to consider pay as an instrument, or for what it is used. The worker's financial commitments, as reported in the interview, may be arranged in four groups. His dependents (wife and children) are the first and most obvious. His home is the next: it may be a house that he owns, a house that he rents, rented rooms, or a Government hostel (for immigrants). The third point to consider is whether or not he has a car. The last is his social commitments, in entertaining at home or elsewhere and in his or his family's leisure pursuits. All four kinds of financial commitments were simply delimited from the interview records and were then examined in relation to the worker's judgement of his pay. It was found that the first commitment only had a significant relation with this judgement. This might be expected. Under the third and fourth heads, the person's financial commitments are optional. He may prefer to spend a substantial amount of his pay on beer and on betting on horse-races (cf. chapter IX), or on social activities in general. He may decide to run a car, usually an old and decrepit type, instead of relying on the grossly overloaded public transport of his urban area. Or he may decide to the contrary, in either field or in both. In any case, the decision would presumably be related to the worker's perception or estimation of priorities in financial commitments, and therefore probably to his rate of pay, but not necessarily to his judgement of the adequacy of this rate. A wide range of individual differences in what the person wants to take on in either field is to be expected, so that a relation between what he has taken on in either and his judgement of his pay is not to be expected. With regard to the worker's housing, the case is different. He must have somewhere to live; but, for a married man, this is likely to be expensive under any conditions. If he owns a house, he has to pay rates and maintenance (or do the latter himself, which is tedious). If he rents a house, he probably pays at least a quarter of his gross income in rent. If he lives in rooms or in a hostel, he probably saves all that he possibly can to get a home of his own. In any case, he is inevitably committed to substantial expense; so that it is unlikely that there would be any relation between the mere kind of expense and his judgement of his pay.

The case is different, however, with regard to the worker's dependents. Whether he has any, depends solely on his decision. But, if he has any, he must support them and, in the nature of the case, he must accept it that supporting them will be by far the largest one of his financial commitments. Once this responsibility is taken on, it cannot be dropped; so that one might expect to find a relation between the degree of responsibility and the worker's judgement of his pay. The relation found is shown in detail in Table 16.

In this case, simple scale values can be assigned to the categories. The values 0, 1 and 2 represent no dependent children, one or two and more than two, respectively. The same values are assigned separately, in the respective cases that the worker does

TABLE 16

Family responsibilities and judgements of pay

The worker is	The worker's judgement is			Scale values		
	Satisfied	Not	Total	Children	Wife	Total
1. Single, or separated from wife, and not responsible for her maintenance	31	3	34	0	0	0
2. Married, no children, or all children earning: and wife also earns	8	3	11	0	1	1
3. Married, no children, or all children earning: wife does not earn	10	3	13	0	2	2
4. Married, one or two dependent children: wife earns	8	2	10	1	1	2
5. Married, one or two dependent children: wife does not earn	16	15	31	1	2	3
6. Married, more than two dependent children: wife earns	3	5	8	2	1	3
7. Married, more than two dependent children: wife does not earn	12	8	20	2	2	4
Total	88	39	127			

not support a wife, that he partly supports his wife and that he wholly supports her. The addition of the two sets of scale values, as shown in Table 16, gives degrees of family responsibility ranging from 0 to 4. The obvious place to divide the table is between degrees 2 and 3, which is between rows 4 and 5. For this division there is a highly significant association ($P < 0{\cdot}001$) between the worker's family responsibilities and his judgement of his pay: workers with less responsibilities tend to be satisfied with their pay and those with more responsibilities tend to be dissatisfied. This generalization holds regardless of nationality or occupational grade, and regardless of age and length of service, except in so far as these are related to the worker's responsibilities, as the workers with most responsibilities are on the whole older men.

It may be said that this finding could hardly be simpler, and therefore hardly requires interpretation. It seems probable that the final judgement of pay is to some extent a quite direct reflection of the worker's perception of what he has to pay for. As some workers put it directly, the pay is, or is not, enough to live on; the only qualification necessary is that this depends partly on the person's major financial responsibilities. This conclusion cannot be compared with the conclusions of other investigators, as simple possibilities of this order appear not to have been considered by industrial psychologists.

In conclusion, the importance of the two general findings must not be exaggerated. As in the analysis of the worker's judgement of his task, the findings are only statistically significant tendencies. Thus the association described above between certain features of the pay system and of the task and the worker's final judgement is true for only 87 of 127 persons (or 68 per cent); and the association between responsibilities and the final judgement is true for only 85 of 127 persons (or 67 per cent). The combined features of the pay system and the task are not associated with family responsibilities, so that a generalization that included both the main findings should be true for more persons than is either finding alone, though not necessarily for many more. Even so, the two findings together would obviously still give only an incomplete explanation of the worker's judgement of his pay. What other factors might influence this judgement, and how they might be related to each other, are questions that must be taken up in chapters VII and VIII.

The two main findings of this chapter, however, can be regarded as a reconstruction of the worker's final judgement which is simple and intelligible, even if it is by no means complete. The worker judges his pay in relation to three sets of differential possibilities: the possibility of simply working for more, at a higher rate, the possibility of getting more simply because he is worth more (that is, of getting more for little or no more effort), and the possibility of having the kind of task which in itself partly compensates for low pay. Lastly, quite independently of this complex set of judgements of the factory situation, the worker judges his pay in relation to how far he has to make it go, outside the factory. There is no evidence that the worker judges his pay simply as an exactly scaled or measured exchange for his services. From this point of view, it is not surprising that various investigators (cf. Viteles, 1954, ch. X) have concluded that the size of the pay packet is not the only important factor in industrial morale. Due allowance must be made, of course, for the limitations of the particular investigation; for the narrow range of evidence as well as for the unreliability of some of the data. Even so, it seems that what may matter for the worker's specific judgement of his pay is not so much the size of the pay packet as how it is made up and what it must be used for. Pay is a complicated and only partly explicit bargain and, under some conditions, it may be the complications and the implicit assumptions that matter.

CHAPTER VI

THE WORKER'S BOSSES AND MATES

So far as they depend on elements in the person's situation at work, the worker's judgements of his task and his pay have been analysed and, to some extent, reconstructed. There remain to be analysed the less important judgements that the worker makes of his bosses and of his workmates. These are quite distinct judgements, both different in kind and independent in fact. However, both presumably depend on what the other persons in question do, so both may be conveniently considered in the one chapter. The worker's judgement of his bosses (cf. chapter III) is a compound of two independent judgements, of his immediate boss and of the factory management as a whole. For the purpose of analysis, it may be necessary to consider the two elements separately: but the reconstruction of the judgements made must be referred to the final inclusive judgement, as it is assumed that this judgement is the one which would usually lead on to action. The worker's judgement of his workmates is the simplest of the four judgements, and can therefore be analysed directly.

Both judgements, then, must be analysed by reference to what the other persons do, whether these are the worker's various bosses or his workmates. The analysis, unfortunately, depended on limited and incomplete information. The most complete information that would have been required would be the observed behaviour, in all relevant situations and over a long period, of the worker's immediate boss, bosses of higher grades and workmates. As noted in chapter II, there was no possibility of systematic observation either of foremen or of groups of persons at work. In consequence, information derived from observed behaviour was used for minor parts of the analysis (as detailed below), but could not be used as a primary source. The obvious alternative source of information was the behaviour of the persons in question, as reported by them in the interview. This information was more generally usable, but was none the less deficient in some respects. The reported behaviour of the worker's immediate boss was known in detail for the bosses (foremen or professional workers) who were fully interviewed, and for a few others (cf. Table 1, in chapter II); but for some bosses there was no usable information, or none at all. The behaviour of members of higher management was known in all cases, but only in the abstract general form of publicly stated management practices and customary policies. Lastly, there was little information about the behaviour of the person's workmates. Because volunteers only were interviewed, it happened frequently that only some, or none at all, of any particular worker's mates were included in the investigation. Besides this, the interview gathered only information about likes and dislikes in general or, at the most, likes and dislikes of named persons; detailed information about the person's behaviour or about his workmates' behaviour, as he saw either, was for the most part beyond the scope of the interview.

It follows that the analysis of the worker's judgements of his bosses and mates is limited to rather speculative conclusions drawn from insufficient evidence, and based on parts only of the whole number of persons investigated. The account set out below relates to factories A to E for the most part (N = 127), and to factory G (N = 62) in some particulars. In most cases, however, it depends on the fractions of these groups (as specified below) for which there was usable information.

THE IMMEDIATE BOSS

The worker's judgement of his immediate boss was defined as before (cf. Table 9, in chapter III), as a simple dichotomy. This definition depended on the worker's statement of who was his immediate boss and the statements made (cf. chapter III) varied considerably. A few workers referred to a leading hand or to a foreman, so called, who was also a worker, as the immediate boss. In the most common case, the foreman in the conventional sense (a supervisor who does not himself engage in the task) was named. In some cases, in factory A only, both the foreman and a professional worker (a chemist) were named as immediate bosses; and again, in some cases, the worker regarded the professional supervisor only as his immediate boss. A consequence of this variety in the reports made was that some workers specified two or more immediate bosses, indiscriminately. Some shift workers worked under different foremen as the shifts rotated. Some workers regarded both the foreman and the professional supervisor (the chemist) as being the immediate boss; and some workers went even beyond this and specified two foremen and two professional workers as their immediate bosses, without distinction. These complications were easily reduced to order in the definition of the judgements made by the assumption that if the worker named more than one immediate boss, his official boss, as designated by the factory management, was the person who mattered. In the analysis of the judgements made, the reduction to order of the complicated cases was not so easy. The only point that need be made in advance of the detailed argument is that the complications were accepted as they were found: the worker's immediate bosses are whoever he says they are, whether he specifies one person or several, and whether or not he specifies the person who is officially designated as his immediate boss.

The first step in the analysis of the workers' judgements of their immediate bosses was therefore to sort out the usable records of interviews with bosses. The total of these records was 35: 12 foremen and 4 working foremen who were fully interviewed, together with 5 foremen who were interviewed only about their task methods and practices, and together with 14 professional supervisors (chemists) who were fully interviewed. The 35 usable records included 24 of 40 groups of workers (in factories A to E) that were distinguished by the official designation of the boss of the group; but these groups included 90 of the 127 workers interviewed. There was thus some usable, if unsystematic, information about the immediate bosses of a majority of the workers.

The second step in the analysis was to classify the methods and practices of the 35 bosses, as they reported them. It was found that the details of the reports fell into 11 categories, and these were divided, by inspection, into two groups, the methods or practices that appeared to be distinctive of the popular and unpopular bosses respectively, where popularity was determined by reference to the appropriate part of the worker's overall judgement: that is, his specific judgement of his immediate boss. The results of this content-analysis are shown in Table 17.

In this table, the first 6 categories are distinctive of the popular bosses and the last 5 of the unpopular bosses. It must be stressed that this division does not depend on exact statistical tests, which were impossible for most of the categories because of the very small observed frequencies. The division depends only on observed tendencies. None the less, it is psychologically at least plausible, if not obvious. The first 6 categories signify that the boss, according to his own report, treats his workers as persons of equal dignity with himself although of different function; that is, that he is 'democratic' or, more accurately, co-operative. The next three categories (7 to 9) signify that the boss treats his workers with suspicion and some hostility, if not with a clear expectation that they will let him down if they possibly can. The significance of the last 2 categories is not so clear, but can be elucidated by reference to the interview records. Category 10 means that the boss studies the worker's personality and qualifications, makes his decision and proceeds to action, without any reference whatever to the worker and his opinions or preferences. Together with categories 7 to 9, this clearly signifies that the boss is authoritarian or, more simply, bossy. Category 11 belongs with these four varieties of authoritarian behaviour because of its derivation from particular circumstances: if every variety of hectic bossing has failed to get the task done, then the boss who has failed descends to his last resort, to do the task himself.

TABLE 17

Methods and practices of 35 immediate bosses

The boss states that what he does is	Frequency
1. To make his workers at home, introduce them to each other	2
2. To explain, demonstrate or show the task	6
3. To get the workers' point of view, as individual human beings	19
4. To lead, not drive; to ask, not order	5
5. To refrain from bullying, standover methods, cursing and swearing	6
6. To balance tasks between workers, give everyone a fair go at the more popular tasks, at overtime work, etc.	3
7. To be firm when necessary	17
8. To study the workers, keep an eye open for bludgers (loafers), and impose order without argument	4
9. To push the workers if necessary	4
10. To allocate tasks on the worker's personality and qualifications	2
11. To do the job himself when necessary	4

It can therefore be assumed that the dichotomy of these 11 categories is, at the least, a hypothesis worth investigating, even though it has no exact statistical support. The appropriate method of investigation is simply to sum the 6 popular and the 5 unpopular methods or practices, separately, and to determine the relation of the sums to the boss's popularity or unpopularity. When this was done, it was found that both summarized groups of methods or practices were significantly related to the workers' judgements of the particular bosses. For the 6 co-operative practices, bosses who report 2 or more tend to be those who are well thought of, and those who report 1 or none tend to those who are poorly thought of ($P < 0{\cdot}001$, for the whole group of 35 bosses: $P < 0{\cdot}01$ for 21 foremen and for 14 professional supervisors, separately). For the 5 authoritarian or bossy practices, it is found that bosses who report 2 or more of these tend to be those who are not well thought of ($P < 0{\cdot}01$, for the whole group of 35 bosses: $P < 0{\cdot}05$ for 21 foremen and for 14 professional supervisors, separately). Finally, it was found that the two sums of methods or practices were significantly negatively associated ($\phi = -0{\cdot}45$); that is, bosses who report that their methods are co-operative tend not to report authoritarian or bossy methods, and vice versa.

It can therefore be concluded that the analysis of the immediate boss's behaviour, as he reports it in the interview, gives a coherent and intelligible result: he is co-operative in his behaviour as a boss or he is authoritarian (or bossy), or he falls somewhere between these two extremes. This classification implies a scale; and a scale was simply made from the elements by subtracting the sum of bossy practices from the sum of co-operative practices. For this scaled sum, it was found that there is a highly significant relation, as shown in Table 18, with the workers' judgements of their bosses ($P = 0{\cdot}002$ for the whole group of 35 bosses: $P < 0{\cdot}01$ for 21

TABLE 18

The classification of immediate bosses

The boss's summed characteristics are (co-operative—bossy)	The workers' majority vote on the boss is		
	He is good	He is not good	Total
2	5	—	5
1	11	4	15
0	3	3	6
−1	1	3	4
−2	—	4	4
−3	—	1	1
Total	20	15	35

foremen, $< 0{\cdot}05$ for 14 professional supervisors). The tendency is that bosses who are well regarded by their workers are those who report at least one more co-operative practice than their total of bossy practices, and the bosses who are poorly regarded by their workers are those who report only as many co-operative as bossy practices, or less. By reference to the categories of Table 17, it can be seen that this is a reasonable and intelligible result. The immediate boss who is approved by his workers is, on the whole, the boss who is co-operative, even if in some respects he is bossy or authoritarian. The immediate boss who is disapproved by his workers is, on the whole, the boss who is bossy or authoritarian as much as he is co-operative, or more so.

However, if this distinction is both intelligible and statistically significant, it has still to be related to the worker's overall judgement of his bosses. The relation between the boss's behaviour, as he reports it, and the worker's judgement of his immediate boss or bosses may be clear. For practical purposes, the worker's judgement of all his bosses together must be considered. The third and last step in the analysis was therefore to relate the classification of immediate bosses to the workers' overall judgements of their bosses. This could be done only approximately, as there were cases in which more than one boss had to be referred to the judgements of the same group of workers. In these cases, as it happened, all the bosses, or a clear majority of them, were in the one or the other classification, co-operative or bossy. The analysis was therefore straightforward, but the results are none the less approximate. A highly significant relation was found ($P < 0{\cdot}001$) between the worker's judgement of all his bosses together and the classification of the immediate bosses from their own reports. Workers whose immediate bosses are co-operative tend to favour their bosses as a whole, and workers whose immediate bosses are authoritarian, or bossy, tend to disfavour their bosses as a whole. This generalization holds for 61 of 90 workers, or 88 per cent, which figures are probably inflated by the approximations noted above. None the less, it can be concluded that the division between co-operative and bossy bosses (as the bosses themselves report their behaviour) has a significant relation with the worker's judgement of all his bosses. This is more true for the younger workers interviewed, and for the workers with relatively short service, than it is for older workers (aged 40 or over) or for the workers with longer service (more than 2 years). Whatever the relation may be, then, it depends on the worker's experience of his immediate boss mostly for the younger and relatively inexperienced workers. Although the difference is significant, it may not need interpretation, as it is merely one of a number of differences, most of which are insignificant. However, it may be related to another difference, that the younger workers are significantly less satisfied with their bosses as a whole. It is possible, then, that the relation between the worker's judgement and the immediate boss's own report of his behaviour is due in part to inexperience and a tendency to be critical and may be reduced by experience or maturity. In other respects, there are no differences: the relation found holds regardless of occupational grades or nationality.

It remains to consider how far the general finding clarifies the worker's judgement of his bosses. The finding must be taken with some reserve, on two grounds. The classification of immediate bosses can be referred only approximately to groups of workers, and these groups include only 90 of the 127 men interviewed in factories A to E. More important than this is the point that the classification itself derives from the workers' opinions of their immediate bosses, so that a significant relation between the two is given in the operations of the analysis. However, the relation found is not a mere tautology. In the first place, it is the worker's final judgement of all his bosses together which is found to be related to the classification of the immediate bosses. Of the large majority of workers who were content with their immediate boss, less than a half were content with all their bosses. The finding therefore refers to both a judgement and a population which are to some extent different from those from which the classification of bosses was derived. Secondly, the finding at the least clarifies the worker's judgement in detail. He may state that his boss is good or bad in specific ways; but the detail of the six acceptable and five unacceptable modes of behaviour of the bosses goes far beyond what any worker reported, besides that it is the best approximation possible to an independent report of the boss's behaviour. It can be taken, then, that the finding does clarify important elements in the worker's judgement. Good bosses are helpful, friendly, not officious or exacting; bad bosses are unfriendly, officious and exacting. It is interesting that this summary dichotomy derives from the bosses' own statements of their behaviour. Of course these statements have been interpreted, but the fact that this simple interpretation is possible suggests that the bosses interviewed gave a reliable account of their behaviour.

The general finding may be compared with the findings of other investigations, although not many are directly relevant. Jackson (1953) reports an experiment which showed that the worker's opinion of his foreman depends primarily on the foreman's behaviour and only secondarily on the worker's own 'personality needs'. No independent assessment was made of the foreman's behaviour, but the finding does generally support the view implied above, that the worker's judgement is concrete and realistic. Viteles (1954, pp. 325–8) reports that foremen who take into account the workers' point of view have workers who are satisfied with their jobs, or of high morale: which concurs generally with the finding noted. A particular comparison may be made with the findings of Argyle, Gardner and Cioffi (1956). These workers found some significant relations between productivity and the foreman's democratic or authoritarian methods, as defined from his own statements. The more productive workers were those whose foremen were democratic. The issue is confused by the very elaborate analytic procedure used to define the foreman's methods, but the final distinction drawn corresponds reasonably to the distinction above, between co-operative and bossy methods, or bosses. If productivity depends to some extent on democratic (or co-operative) supervision, then it is reasonable to expect that the worker's judgement of his bosses (which precedes the actions that are defined as productivity by at least one stage in the series of behaviour) will also depend to some extent on the nature and quality of supervision. The general finding, then, may be taken as conventional, although the detail of the categories that constitute co-operative or bossy behaviour is an addition to previous findings.

The conclusion, therefore, is that the worker is satisfied with his bosses as a whole if his immediate boss behaves towards him as a person of equal dignity, though with different and more complex responsibilities; and is dissatisfied if his immediate boss behaves as if he were a superior person, by virtue of his place and function. While this conclusion depends on a merely significant association, it does show that the worker makes his final judgement of his bosses by detailed, realistic and direct reference to the overt behaviour of his immediate boss. His final judgement obviously depends on other factors, both within the factory and outside it, but it does depend on this recurrent daily experience to a significant extent.

THE HIGHER MANAGEMENT

The next step in the analysis is to consider the relation of the worker's judgement of his bosses to the behaviour of bosses of higher grades. There was considerable variety among the workers interviewed with regard to the bosses, and hierarchical grades, on whom or on which they could make specific comments. At one extreme, a few men knew no boss except the leading hand (or working supervisor) of their section. At the other extreme, a few men were able to name and describe as individuals a number of bosses in grades from just above the foreman to the board of directors. The variety between these extremes is reducible to order by reference to information of two kinds.

For bosses of intermediate grades (junior or senior executives, but not managers or directors) there are numerous interview records, but few of these are usable for the present purpose. As a matter of policy, all higher bosses were interviewed only as expert consultants. Most of the records therefore include the person's general opinions on policy and hypothetical practices rather than direct statements of behaviour. Besides this, most of the intermediate bosses interviewed had little or nothing to do with the workers directly. There was an exception in factory G. In this factory, there were five general foremen, or supervisors, whose official place in the hierarchy was distinctly above that of the first-line foremen (or non-working supervisors, in the usual sense). The official status of these general foremen was understood by most of the 62 men who were interviewed in this factory. However, these five men were not executive members of management: they were promoted workers who wore the same sort of dust-coats as their foremen, and who were paid no very substantial margin above the rates for foremen in the ordinary sense. All five were interviewed, as expert consultants, and all five expressed themselves freely and at some length, in spite of the fact (or perhaps, by reaction, because of the fact) that the investigation in this factory only was done on behalf of the management. Besides this, all five were extensively observed at work. No attempt was made at systematic observation; on the contrary, the particular circumstances of the investigation in this factory excluded deliberate observation even more firmly than in the other factories. But all five general foremen had extensive territories (they had charge of 80 men, in the least case, to 400 men at the most) and all five regarded their job as requiring them to supervise their foremen, or even their workers directly, as their chief responsibility. In consequence, all five spent most of their time on the factory floor and were therefore observed in action, incidentally and undeliberately, but frequently.

The case with members of top management is different. All of these, whether managing directors, factory managers, assistant managers or personnel managers, were interviewed only as expert consultants. All therefore gave considerable information (reviewed in chapter X) about the policies and practices of the firm, in which information on personnel policies and practices was specially relevant. It was therefore easy to put together a summary statement for each firm (factories A to E, and G) showing the distinctive features both of general policy and of personnel policy. In consequence, it was possible to refer the workers' judgements of their bosses to a classification of these six factories that was based on differences in policies and practices.

The particular case that may be considered first is the relation of the worker's judgement of his bosses to the behaviour of the five general foremen in factory G. The relevant judgement is the judgement of all bosses together, as previously, because this is the judgement which is likeliest to lead on to action in the factory situation. The behaviour of the five general foremen can be assessed from interview records, backed up by incidental but extensive observation, as noted above. The analysis of this evidence showed that two of the general foremen could be classified as co-operative and three as officious or bossy. The distinction was not entirely clear from the interview records, but was clarified beyond doubt by observation. This point is worth stressing, because of the paradoxical nature of the findings that are set out below. One general foreman, who had apparently decided that he ought to vindicate his methods to the investigators, took the author on a tour of his extensive territory just after he had been interviewed.

In the course of this tour, he demonstrated the difficulties of his task by dragging a woman away from the machine she was operating and vehemently denouncing her, on the spot, as an intolerably incompetent worker; while, with equal vehemence, he pointed to the products she was working on and deplored their defects. The status as evidence of one incident may be doubted; but this incident was merely a dramatic example of a number of similar though less dramatic incidents that were observed. Further, all these incidents were generally in line with the interview record for this particular general foreman. Another of the general foremen had apparently made it his business to detect and reprove any idleness that he observed in the course of his movements through the factory, even in the case of workers who were beyond his control and outside his physical territory. Shortly before he was interviewed, this general foreman took occasion to reprove a tradesman whom he noticed apparently doing nothing. The tradesman took the view that the foreman had no authority whatever over him and should therefore mind his own business; and the end of a sharp dispute was that the foreman, who was much the smaller man, was prostrate and unconscious. An unfortunate consequence of this incident was that none of the group that the tradesman belonged to were willing to be interviewed, although several had shown keen interest before the incident. It must be made clear that this incident was not observed. It was reported, however, in almost identical terms by four workers who stated that they had seen it. More importantly, it was reported in interview, in substantially the same terms, by the general foreman himself, who quite cheerfully asserted that he had been unlucky on that particular occasion but that he had done no more than his duty and would do it again. It may be added that this person also detailed his activities as the firm's self-appointed detective, the extreme case of which was his driving in to the factory from his distant home, in the small hours of the morning and of his own volition, in order to apprehend in the act a despatch driver whom he suspected of stealing the firm's products.

These two dramatic incidents may have little status by themselves as evidence; but, in so far as they are backed up by other less dramatic observations and by the general foreman's own statement of his behaviour, they must be taken as conclusive. Three of the five general foremen in the factory G were officious or bossy in a degree far beyond the most extreme case observed among the immediate bosses in factories A to E; and that these three were so was visible to the most casual observer. The other two were relatively co-operative, as this was determined from the interview records and observations; though, as the factory situation at that time favoured bossy bosses, neither was as openly and clearly co-operative as the immediate bosses in factories A to E.

Once the classification of the five general foremen is established, the next move is to relate this to the judgements of their bosses as a whole that are made by workers who comment specifically on one or more of the general foremen. When this relation was made, the results were contradictory. Most of the workers concerned were in the charge of the general foremen and for these workers it was found that the positive judgement of the bosses as a whole was significantly associated ($P = 0{\cdot}003$) with the person's working for an authoritarian, officious or bossy boss. A minority of the workers who made comments on the behaviour of the general foremen were maintenance tradesmen who noted this behaviour as relatively independent observers, in the course of their duties in the territories of, but not in the charge of, the general foremen. For these workers it was found that their judgement of the bosses as a whole was not significantly associated with the classification of the general foremen as co-operative or bossy. This is not remarkable, as these men were not their bosses in any official sense. It was also found that these relatively detached observers took a significantly worse view ($P < 0{\cdot}0001$) of the five general foremen than did the men who worked for these five.

This finding gives the clue to the puzzle. It is a grave matter for any investigator to propose that his own observations should take precedence of field findings in general, especially if the observations are not systematic enough to be capable of exact scientific analysis. However, in the particular case, there was no possible doubt whatever that the investigator's classification of the five general foremen was correct, as their behaviour was gross, obvious and directly visible in many instances, besides that their own statements made in interview largely concurred with inferences made from observation. It can be concluded, then, that the tradesmen who judged these five general foremen from a relatively detached point of view judged them realistically, while the men who worked for these foremen did not do so. The difference is not surprising. As the investigation was introduced in Factory G (cf. chapter II), nearly all of the process workers and labourers who were interviewed were the foremen's or general foremen's nominees and not volunteers. On the other hand, a majority of the maintenance tradesmen who were interviewed (20 of 62 men altogether) were men who were approached individually by the investigators in the course of the field-work. It can therefore be concluded that the difference noted is due only to the conditions of the investigation in the particular factory. The observed fact that some workers positively like bossy bosses cannot be taken to upset the general conclusion stated above, that co-operative bosses are on the whole the more liked.

The fact is none the less interesting. It may be that only the general foremen's nominees are well taken with the singularly officious or bossy bosses of factory G. This possibility is supported by the observation that the process workers and labourers interviewed in this factory had much longer service than the tradesmen; among the former, over 20 years' service was not uncommon, but was unknown among the latter. Even so, it is still interesting that there are any workers at all who like officious or bossy bosses, as three of the five general foreman in factory G undoubtedly were. There are two possible explanations. There may be some workers whom it suits to be kicked by the boss, or who at least get used to it after the first few years. Alternatively, it may be that the officious or bossy boss reserves his active officiousness for the relatively junior workers; so that, if the worker has held his job for ten years, he has earned his place and is trusted and let alone, rather than kicked. This possibility is to some extent supported by the observation that some independent tasks (cf. chapter IV) can be done only by a worker of long experience. At all events, it is the likelier possibility, as it depends only on elements in social psychology, while the first possibility would require an interpretation in terms of the darker parts of psychopathology. Whichever explanation is adopted, the general conclusion remains, with one modification. In general, workers prefer co-operative to officious bosses: but this tendency may be reversed in certain circumstances which cannot be clarified from the information available.

The second case to be considered is the relation of the workers' judgements of their bosses to the behaviour of the top-level management as a whole. This behaviour was not observed, not only because it could not be observed within the practical limits of the investigation, but also because it was for the most part not capable of observation, in its nature. General policies and personnel practices, as defined by the factory management, may be related to the worker's judgement of his bosses. Policies and practices were adequately stated in the interviews with members of higher management. It was not practicable to take into account the effectiveness of these policies and practices, the estimation of which would have required far more evidence than was available. Nor was it necessary to consider effectiveness in this sense. The relation between the worker's judgement of his bosses and his firm's officially defined policy and practices may be investigated by itself, as the official definition presumably bears a close relation to practice. In consequence, the records of interviews with senior executives and managers or directors were used only to establish a classification of the six factories A to E and G. Apart from minor differences among factories A to E (which are reviewed in chapter X), the main division established was between these five factories and factory G. In factories A to E the top management had some effective contact with the workers and tended on the whole to co-operative policies; while in factory G the top management was remote from the workers and tended on the whole to officious and bossy practices. This classification depends, again, on the investigators' interpretation. It is doubtful whether it could be supported by the exact analysis of interview records but, as with the general foremen in factory G, the difference was gross, obvious and immediately visible. It may therefore be accepted, as a starting point for the analysis of the workers' judgements of their bosses, on the ground of common experience rather than the ground of scientific demonstration.

It was found that the workers interviewed in factory G (62 of 189) were significantly less content with their bosses as a whole ($P < 0{\cdot}05$) than were the workers interviewed in factories A to E (127 of 189). This finding was clarified by reference to the workers' specific judgements of the top management. It was found that workers in factories A to E (and especially in A alone, in which 58 men were interviewed) tend to think well of the management in general, whereas workers in factory G tend to a poor opinion or to regard the top management as unknown ($P = 0{\cdot}03$). The analysis of unfavourable judgements of the management in general extended this finding. It was found that workers in factory G tended either to state that they had no knowledge of the top management or to criticise it drastically: while workers in factories A to E tended to make specific critical comments that were limited to particular members of the top management. It may be concluded that there is some relation between the worker's judgement of his bosses and the general policies or practices of the firm. The relation is clear in one case only. The origin of the distinctive features in this case (factory G) must be left for further discussion (cf. chapter X). For the present purpose, it is sufficient to make the point that, even within the limits of the investigation which was carried out only in relatively progressive factories, the worker's judgement of his bosses does depend on the factory's general policy and practices. The relation is as before: workers who think well of their bosses are those whose top bosses are known and liked as persons, while workers who think poorly of their bosses are those whose top bosses are either unknown because they are too remote to be known or disliked because they are officious or bossy.

It can be concluded that the worker's judgement of his bosses of higher grades, up to the top management, is made in much the same terms as his judgement of his immediate boss, only with less evidence at his disposal. It is a concrete and realistic judgement, that depends on the same general distinction between the observed behaviour of co-operative and officious, or bossy, bosses. It seems, therefore, that this is the distinctive ground of experience which underlies the worker's judgement of all his bosses together. However, two qualifications must be made. This generalization is certainly not true for some workers: on the contrary, in some cases the reverse relation holds. It may be suggested that what the boss does depends on with whom he does it; that is, that the apparent uniformity of behaviour inferred from interview records masks an actual diversity of behaviour which is related to differences in standing and function among any one boss's workers. A more important point is that the general relation found, if it is coherent and intelligible, is far from complete: it is clear that there must be weighty elements in the worker's judgement of his bosses beyond his recurrent experience of their behaviour. This question, however, must be left for consideration in chapters VII and VIII.

THE WORKER'S MATES

The information available is obviously insufficient to determine the grounds of the worker's judgement of his general and continuing relations with his workmates. This judgement may or may not be realistic and coherent; it may be a considered reflection of the details of his relations with his mates or it may be a conventional stereotype, or a reflection of what he would like to do or thinks he ought to do. What he actually does with his mates presumably depends both on what he would like to do, being as he is, and on what he can do in the factory situation. The possibilities of the latter depend, again, both on the range of social contact open to him on the task and off duty, and the kind of particular contacts open, or the particular individuals with whom he must or may associate. The significance of the worker's being the kind of person he is (or of his 'personality') must be set aside for later consideration (in chapters VII to IX). For the other factors, the information is very limited. The interview records specify with how many others each man works, but it was not possible to go beyond this bare statement and to infer from observation what degree of functional dependence the task required, or what amount of conversation it permitted. Similarly, the interview records give some detail, but not enough, of the worker's general relations with his mates. Beyond his general statement of how he gets on with them there is, in some cases only, detail of particular persons liked or disliked and, in even fewer cases, detail of particular common activities (such as playing cards in the lunch break) or particular reasons for dislikes. It was thought undesirable, however, to press the interview beyond this insufficient level of detail, both because of lack of time and because the men interviewed might have been disinclined to go into more detail. A further restriction is that where the person does refer to other persons by name, these may be men who were not interviewed. The consequence of these various deficiencies, which were inevitable in the nature of the project, is that there is some detail, though not enough, for 53 or less of the 127 men interviewed in factories A to E; for the majority, there is no usable information at all. Meagre as it is, the information may be worth analysing.

The analysis of the worker's range of social contacts is easily disposed of. No relation was found between the number of persons the man says he works with and his judgement of his mates. There is no further detail of the task requirements and conditions that might be relevant, and those that were reviewed in chapter IV can be assumed to be irrelevant. For 127 workers, then, it is found that the worker can still be sociable if his task requires him to work alone: there are plenty of other occasions for being sociable during the working day. Or, at the other extreme, the worker can still be unsociable even if his task requires him to work in a gang. Presumably he will get by if he does his work reasonably, but engages in the bare minimum of personal relations necessary for this. It cannot be said, of course, that the worker's judgement of his mates does not derive at all from the conditions under which he and they work, but it seems likely that the worker's social behaviour, and therefore his judgement, could be independent of task conditions over a wide range.

The next point to consider is the worker's specific comments about other workers. These were classified by the conventions of sociometry: triangular choices, paired choices, single choices, doubtful choices (all of which were single), single rejections and paired rejections. No other types were found, which by itself shows the incompleteness of the data; and for this reason there is no point in detailing the number of each type of comment. A further complication was that the information varied among the 53 men for whom there was some: 21 only of these both named particular workmates and were named (as approved or disapproved) by others; another 17 were named but themselves named no individual who was interviewed, and the remaining 15 made comments on individuals but had none made about them. The obvious way of reducing this confusion to some sort of order was to sum both the choices made and the choices received by allotting weights from + 3 to − 2 to the six types of comment set out above, and adding. The two totals were then tested against the worker's general statement of how he gets on with his mates.

The first test, the details of which are set out in Table 19, showed that there is a highly significant relation between the worker's general judgement of his mates and their opinion of him, where this is expressed as the sum of the scale described above and a sum of + 1 or more signifies approval while a negative or zero sum signifies disapproval. The relation holds for all 38 men on whom specific comments were made, and is somewhat more marked for the 21 men who also make comments (not, in all cases, mutual in nature). This is an obvious finding. By itself, it means only that the men who say that they get on with their mates in

TABLE 19

The worker's general judgement of his mates and their comments on him

	The sum of his mates' comments is	The worker's general relations with his mates are		
		Good	Not good	Total
(For all men on whom comments were made)	1 +	17	8	25
	0, –	2	11	13
	Total	19	19	38
(For all men who also made comments)	1 +	9	3	12
	0, –	—	9	9
	Total	9	12	21

general tend to be those whose mates say that they get on with them in particular. However, this is worth knowing, as it suggests that the general judgements made by 38 men at least could have been backed up by realistic and verifiable detail; that is, that they were not merely conventional stereotypes. This is supported by the second test, of the worker's general judgement against the scaled sum of his own comments on his mates. Here again, a significant relation was found, although it was less marked: the workers, whether 38 or 21, who speak well of their mates in general, tend to be those who also speak well of their mates in particular. As this is an accessory finding, the details need not be tabulated, but a third finding may be added: that the two scale sums of comments, made about the worker and made by him, are highly associated ($\phi = 0{\cdot}81$, for 21 cases). Both these findings strongly support the suggestion that the worker's general judgement is considered and articulate and not merely a conventional stereotype.

This is as far as the analysis can be taken. It is worth showing that the judgements of some workers, at all events, are considered and articulate, but it is not possible to say on what kinds of behaviour or expectations they are based. There is so little detail of who does what with whom that one cannot construct a picture of approved or disapproved kinds of behaviour, let alone of the reasons for approval or disapproval. A few general suggestions may be made, from the few really detailed interview records. It seems that a close mate may be one with whom the worker talks, jokes, exchanges confidences, plays cards and so on, but does not usually see outside the factory. Only 2 of 127 workers were on terms of close friendship with workmates and only 50 had any acquaintance at all with their mates outside the factory; though this, of course, may be due in part to the sheer size of Melbourne, which strongly favours neighbourhood friendships. A mate with whom one merely gets on may be someone whom one has little to do with, but that little on conventionally friendly terms. A worker with whom one does not get on seems likely to be an individualist: a chronic grouser or loafer (at the expense of his fellow workers) or an eccentric who parades his views. There were three or four such persons only who were clearly not on good terms with their fellow workers, but in no case was there any serious strife. In all cases it seemed quite possible for all the workers concerned to get on with their work and for the individualist to get by, which presumably only reflects the low order of importance of the worker's general judgement of his mates.

In conclusion, it cannot be said that this judgement does not reflect the local situation, whether that is the conditions of the task or the behaviour of the worker's mates. It can only be said that the judgement is specific, articulate and considered; and even this holds definitely for only a small fraction of the total number of men interviewed. It would doubtless hold for some others, who could not be included in the analysis merely because the men about whom they made detailed comments had not been interviewed. But there were other men, again, whose sole comment on their mates was a brief and general stereotype. The basis of such judgements could be determined only by the analysis of the actual relations of persons at work, on which almost no information was gathered.

However, it could be argued that, if the worker's judgement of his mates is a realistic reflection of their behaviour towards him, this is the only substantial connection that there is likely to be between the local situation as it would be seen by an observer, and the worker's judgement. Of course this connection would still require great clarification in detail; but the alternatives in the local situation may not be of great moment. The most obvious alternative is that the behaviour of the worker's mates, as he judges it (and, of course, his behaviour as they judge that) is determined in relevant respects by the conditions of the task. It has already been suggested that this is unlikely as a general case, because few tasks appear to define incidental social behaviour stringently (even though a few exclude any social behaviour) and even more because there is ample scope for unprescribed social behaviour off duty. It is true that Trist and Bamforth (1951) have described an interesting case of serious hostility between groups of workers that was derived from the conditions of the task; but this case can hardly be regarded as typical of industry, especially because of the very inept management to which the objectionable task conditions were due. The less obvious alternative in the local situation is that the worker's judgement depends on the luck of the draw. If he is socially inclined and finds himself working with a gang of morose neurotics, he may or may not get on well with them, and so on. Whatever he does, he is still as likely to make a realistic judgement of their behaviour as he would in a more ordinary situation. Just how this judgement derives in detail, from the local situation, would then require clarification, as for any other judgement. But this possibility would not be an alternative in the local situation unless the general finding, above, that sociability and unsociability are on the whole reciprocated were found not to hold in this case. It would be interesting to know under what conditions reciprocity of judgement does not hold. Until this can be answered, the question remains, how did it come about that the one man is socially inclined and the other men are morose neurotics and so on. This question belongs to their personal history, and not at all to their present work situation; and an answer to it will be attempted in chapters VII to IX. The point, meantime, is that this is perhaps as likely a way of analysing or reconstructing the worker's judgement of his mates as is the unfortunately impossible detailed analysis of the local situation.

It is clear that the findings of this chapter must be taken with even greater reserve than those of the preceding chapters. The explanation of the worker's judgement of his bosses, as set out above, holds for 61 of 90 workers, or 68 per cent. Besides that this is a significant tendency but by no means a complete explanation (as with the worker's judgements of his task and his pay), this finding derives from a part only of the total number of men interviewed, and its derivation is approximate because of the uncertainties in the link between the individual worker and the immediate bosses who were interviewed. It follows that the findings about the worker's judgement of bosses of higher grades and of the top management are even more speculative. Lastly, there is no finding about the grounds of the worker's judgement of his mates. The most that can be asserted about this is that it is a coherent and considered judgement; but even this is known to hold for only a minority of the men interviewed.

Even so, the findings are reasonable, as far as they go. It is, at the least, a hypothesis worth pursuing that the worker's judgement of his bosses depends partly on whether their behaviour is co-operative or despotic. It follows that this judgement may also depend, in part, on the worker's frame of reference, or on his experience in general or on his particular experience of different bosses. In the same way, the worker's judgement of his mates, which apparently depends little on his actual situation at work, may depend on various elements of his experience. The same might be true for his judgements of his task and pay, both of which depend significantly, but only in part, on identifiable elements in his work situation. The next step in the argument, therefore, is to consider the worker outside the work situation, in his wider context as a member of an industrial society and as a person with a particular kind of history.

CHAPTER VII

THE WORKER'S PLACE IN SOCIETY

THREE significant elements in the worker's judgement of his situation at work have been discussed so far. It appears that the worker's final judgement of his task may depend on the task's being clean, or independent or complete, or on some combination of these: at all events, the judgement made is made to some extent by reference to the task itself, or the worker's present situation at work. In the same way, the worker's judgement of his pay may depend on whether or not he is paid something (of whatever kind) above the local official base rate, for little or no extra individual effort; and his judgement of his bosses in general may depend on whether his immediate boss is co-operative or bossy, as he himself describes his behaviour. These two judgements, equally, are to some extent made by reference to the work situation directly, whether to the pay system or to the behaviour of the immediate boss. The same cannot be stated for the worker's judgement of his fellow-workers; this may depend on the detailed behaviour of the person's workmates, but the evidence is insufficient to establish any such dependence. Lastly, the worker's judgement of his pay may depend also on his judgement of what it must be used for (i.e. the number of his dependents); and this judgement is of one element in his whole situation or set of engagements as a person, but is not a judgement of his work situation at all.

However, the worker's judgements of his situation at work derive from elements in this situation to a limited extent only: all three of the findings are only partial and approximate reconstructions of the way in which the worker makes his final judgements of his work. It is therefore to be expected that these judgements may also derive to some extent from the worker's judgements of situations other than work. The second finding for pay shows this, in a particular case; the worker's judgement of his pay may depend as much on what he has to use it for as on what it is given for. The general possibility, then, is that this judgement and the others may depend on various elements in the worker's whole set of personal engagements, or his judgements of these. The possibilities are limited, in the actual case, by the information available, but four may none the less be distinguished. The person's present engagements other than work, or his judgement of these, may be relevant: a possibility which is examined in chapter IX. Equally, the person's habitual or recurrent relations with other persons in general, or his general judgements of other persons in relation to himself, may be relevant; and so may be his general judgement of his personal future prospects in relation to his past. These two possibilities are examined in chapter VIII, so far as there is information about them. Lastly, the person's place in his society, or his judgement of this, may be relevant; which is the matter for this chapter.

It appears that there are some workers who regard their task, pay or bosses as satisfactory in spite of the particular elements of their work situation; that is, even if the task is dirty or dependent or incomplete or some combination of these, even if their pay includes no notable extras, and even if their immediate boss is officious, bossy or exacting. Further, these workers are a considerable minority in all three cases. The same holds for the converse cases generally. There are some workers who regard their task as unsatisfactory even though it is clean, independent or complete, or some combination of these; some who dislike their pay even though it includes substantial extras, and some who dislike their bosses even though their immediate boss is co-operative. These again are a minority, but a considerable minority, in all three cases. A considerable minority of workers, therefore, make judgements of their situation at work which apparently do not depend at all on this situation, as they judge it or as it is judged by the detached observer. Among the various possibilities, the one that must now be considered is that the judgements made by these workers (and maybe by other workers equally) depend on the worker's place in his society, as he sees it. The worker's task is a present fact, and he may have undertaken it deliberately, in so far as he could. That is, he may have done what he could to find himself a task that has one or more of the features that are apparently desirable. But what any worker can do in this sense must be limited by his circumstances. It is obvious that no person can abstractly consider the whole range of tasks that men are paid to do, and then make a considered decision for just the one task that he prefers. On the contrary, the possible range of decision is narrowly limited, both by the worker's limited qualifications and by his ignorance of the possibilities even within the range of tasks for which he is formally qualified. It follows that some workers may do tasks which, as they see them, are not the sort of tasks that they ought to be doing or really like to do. The settled worker, at least, is fixed once for all as a particular sort of tradesman or as a process worker or a labourer; but he may not have wished to have his occupation fixed, even in this relatively general sense. It follows that his specific task may be one that he regards as quite appropriate, or it may be one that he regards as somewhat beneath him, or even heartily despises, as being out of line with his proper place in society. The general possibility is the same for the worker's pay, bosses and mates.

The possibility must be referred, of course, to the worker's place in society as he sees that, and not merely to his place as the detached observer sees it. The latter is given: all the men in question (in this chapter, 127 men who were interviewed in factories A to E) were manual workers, who were sub-classified as tradesmen, process workers or labourers. No differences were found between persons in these sub-classes in any of their judgements of the work situation. But this classification does not, of itself, give any indication of the worker's judgement of his place in society. It is not the fact that must be considered, but the person's evaluation of the fact; whether or not he is content with his place in society, on what basis he makes his judgements of this place, and the relation of these judgements to his judgements of the work situation. In the interview (in factories A to E only) information was gathered about the person's general view of social classes, his political and trade union activities and opinions, his detailed views on social classes, and his occupational history. The definition of his judgement of his place in society, as he sees it, may depend on any or all of these; and all must therefore be examined in turn.

THE WORKER'S POLITICS

The most general expression of the worker's view of his place in society presumably is his political opinions and practices. If he is content to be a manual worker in general, he will tend to be conservative; if not, he will tend to be radical. This may be referred, firstly, to the worker's general opinions of the social classes in Australia, as set out in Table 20. If he is content to be a manual worker, he should prefer to describe his society in hierarchical terms or not in terms of classes at all (categories 1, 2 and 5 in Table 20); while, if he is not content to be a manual worker he should prefer the description in terms of conflict between classes (category 4). The table shows a summary of the classifications stated, together with a summary of those reported by Hammond (in Oeser and Hammond, 1954, p. 273) for a small but carefully selected general sample of manual workers. The two classifications do not differ significantly in any respect. It can therefore be taken that the opinions of 127 workers interviewed were as representative of the general working population as were those of Hammond's sample. However, it was also found that there is no relation whatever between these opinions and any of the worker's judgements of his work situation. The worker may or may not be satisfied with his task, pay, bosses or mates, regardless of his general view of the structure of society. It may be that this view is too general to reflect his judgement of his own place in

TABLE 20

Definitions of the social classes in Australia

Classes named	Incidence in	
	This investigation	Oeser and Hammond
1. Rich, middle, poor	11	1
2. Upper, middle, lower	25	7
3. Upper, middle, working	59	32
4. Capitalists, workers	21	9
5. No social classes at all	11	6
Total	127	55

society, or it may be that a general view of this kind is, for relatively unsophisticated persons, a forced and artificial statement of unconsidered opinions. At all events, the general view appears to play no part in the worker's judgement of his situation at work.

TABLE 21

Distribution of political votes

The worker votes for	Incidence in	
	This investigation	Oeser and Hammond
1. The Labour Party	89	31
2. The Liberal (i.e. Conservative) Party	13	7
3. Either, depending: undecided: not disclosed	23	1
4. The Communist Party	2	3
Total	127	42

The next step, then, is to consider the direct expression of the worker's political orientation; his party, as shown in Table 21. In this table, it appears that the political affiliations reported differ significantly ($P = 0{\cdot}02$) from those reported by Hammond (in Oeser and Hammond, 1954, p. 307) for a sample of manual workers, in that there are more swinging or undecided votes among the 127 men now under consideration. This difference does not appear to be of any psychological consequence, especially as no relation was found between the worker's political affiliations and any of his judgements of his situation at work. In this case, it seems likely that the worker's decisions in voting are too simply grounded to bear any relation to his somewhat remote judgements of his work. In Australia, everyone is required by law to vote in elections for both the upper and the lower houses of both the State and the Federal parliaments, so that the person has to vote in some election about once a year. In these circumstances, it may well be that many persons regard voting merely as a routine chore, which may have a roundabout and hypothetical relation with their concrete personal engagements in the future, but which is not likely to produce any predictable or obvious consequences in these engagements. This speculation is supported by the opinions of politics in general that were stated by the workers interviewed. A small minority (15 of 127) stated that politics were important, should be studied by the voter, or should be taken seriously. A larger minority (39 of 127) were uninterested in politics to the extent of saying so, briefly. The majority (73 of 127) inclined to the view that politics in Australia are a self-contained field of doubtfully honest activity that bears no relation at all to the interests or purposes of the ordinary citizen; and, of these, 63 (or half of the whole group) expressed this view ('all politicians are crooks', etc.) without any favourable qualification at all. While no direct comparison with other findings is possible, it seems likely on general grounds that these findings are not grossly unrepresentative. It may be concluded, then, that political activity is not of great interest to many Australian workers; so that its lack of relation to the worker's judgement of his work situation is not surprising.

The case might be different, however, for the worker's trade union membership and activities. It may be that the analysis has been concerned so far with activities or opinions that are too distant from the worker's immediate situation to enter into his judgement of this. But his participation in trade union activities is likely to be on immediate grounds to some extent. Membership may be partly determined by ideological, or general political, considerations, as well as by the exigencies of the situation (such as factory policy). However, the worker's activity, beyond merely paying his dues, presumably depends to some extent on what he expects to get from the union in the satisfaction of his present needs in the work situation. Among the 127 men interviewed there was sufficient variety of reported behaviour for this possible relation to be tested. Most of these men were trade unionists (81 of 127, which is a high figure if one takes into account that one of the five factories A to E was an open shop and another a non-union shop, and that a number of workers who were quite willing to be members were equally willing to leave it to the union to take the first step). Of this majority, 7 only were union officials (not all of whom were active in union affairs, in the factory), but 17 altogether (4 officials and 12 ordinary members) stated that they attended union meetings regularly. The remainder were passive members who merely paid their dues. While no exact comparison is possible, this distribution of union affiliations seems to agree generally with Hammond's brief remarks (in Oeser and Hammond, 1954, pp. 21, 22) on the same subject. At all events, there is sufficient variety, from active members through to non-members, for the worker's trade union affiliations to be capable of relation to his judgements of his work situation. When this was tested, however, it was again found that no one of the four judgements had any significant relation with any type or degree of union affiliation.

As with the worker's political activities, it seems likely that this negative finding is due to the remoteness of the behaviour considered from the worker's immediate interests in the work situation. It is true that some workers appreciate the material gains that they have derived from union activity: 31 of 127 mentioned particular gains in wages, hours of work or amenities. It is also true that a half had some doubts about trade unions in general: 54 of 127 thought that they were a good thing, 9 disapproved of unions generally, 8 had no clear opinion and 56 gave only qualified approval. Finally, it is notable that the tone or quality of the critical remarks made by these 56 workers was much the same as for the critical remarks made about politics: unions are all right, or useful, if they are properly run, if they do not go on strike for nothing, if there are no ratbags in control, and so on. It may be concluded that some workers, at least, regard their trade unions as sceptically as they do politicians; so that, in spite of the union's ostensible concern with the worker's immediate needs, it is not surprising that there is no relation between the worker's judgement of these needs and his union affiliations.

It may be recalled that the general class opinions and political affiliations of the men in question differ trivially or not at all from those of the small but carefully selected general sample with which they were compared in Tables 20 and 21. There is, therefore, some ground for supposing that the men interviewed were not grossly eccentric in their views. In consequence, it seems likely that the definition of the worker's place in society from activities and opinions which can all be called political in a broad sense fails because these activities and opinions are too remote from the worker's concrete daily engagements. Simply as a definition, any one of the sets of data analysed would be perfectly valid; but all appear to be definitions of judgements that have no consequences in the worker's immediate judgements of his work situation. There remain to be considered other possible definitions of the worker's place in society, as he sees it.

THE WORKER'S SOCIAL CLASS

From the observer's point of view, all the men interviewed were manual workers, or members of the working class in general, and were tradesmen, process workers or labourers specifically. The worker's own point of view may differ from this, but its definition must depend on the procedure used to elicit it. For this part only of the interview (points 34 to 38 in the check list shown in the Appendix), a fixed order of questions and a fixed form of words were adopted, on the advice of the investigators' colleagues (cf. Hammond, in Oeser and Hammond, 1954, pp. 265–8). The first question therefore elicited the worker's view of social classes in general, as summarized in Table 20. For the majority, who mentioned two or three distinct social classes (in their own nomenclature), this was sufficient or almost sufficient. A minority, however, described the structure of Australian society by reference to the diversity of social, sporting and religious organizations, etc. Statements in these terms were accepted meantime and were recorded as they were made.

The worker was then asked to name his own class. All those who had begun by describing their society in terms of social classes placed themselves in one of the classes they had previously named; and some of the minority who apparently saw no social class structure in their society none the less placed themselves in a named social class. At this stage, the specific prompt-question was asked of the remaining minority who had made no reference to social classes in the ordinary sense: 'Some people refer to the classes as "working class", "middle class" and "upper class". Which one would you say you belonged to?' The responses to this direct question reduced but did not eliminate the minority who were not prepared to place themselves in a social class. The worker was then asked what he thought in general of the classes specified, from highest to lowest in conventional terms. Where he had himself named social classes, the question was put in his own terms. For the minority who had not done so, the question was put in terms of the direct prompt: upper, middle and working. This was done for the whole minority, whether or not they had classified themselves in response to the direct prompt. Lastly, the worker was asked to class persons in ten specified occupations that were presented to him on a card (see point 37 of the check list, in the Appendix). In this case, it was left to him to refer back to his previously stated framework, of whatever kind, or to use another framework, whichever he preferred.

This whole set of information may or may not be related to the observer's definition of the worker's class or sub-class. The first point of interest is the definition of the worker's own view of his social class, which can be attempted in several ways.

The first and most direct of these is from the worker's statement of his own class, as summarized in Table 22 in relation to similar results from certain other sources. In this table, the categories are defined thus:

1. The worker is a member of the working class if he says so, using this term; or if he places himself in the lowest of his stated two, three or four classes, excepting only if the lowest is named with an emotional tone (e.g. the slum class, ratbags, deadbeats, no-hopers).

TABLE 22

The worker's statement of his own class

The worker states that	Percentage incidence in		
	This investigation	Oeser and Hammond	Centers
1. He is working class	72	64	77
2. He is middle class	23	25	21
3. There are no social classes *or*, he cannot place himself	5	11	2
Number of cases	127	55	414

2. The worker is a member of the middle class if he says so, using this term; or if he places himself in the upper of two named classes or the middle of three, excepting only if his lowest stated class has an emotional tone, as above. It may be noted that no workers placed themselves in a lowest class of this sort, or in an upper class of three stated, which latter therefore does not require definition.

3. The worker is unclassified if he refuses to classify himself, either because he finally repudiates the idea of social classes or because, while he accepts it, he cannot make up his mind where he himself belongs.

It is notable that a large majority of workers placed themselves where the observer equally would have placed them. It is also notable that the distribution found agrees closely with that reported by Hammond (in Oeser and Hammond, 1954, p. 268) for a smaller but carefully selected general sample of working men; and also agrees closely with that reported by Centers (1949, p. 86) for a large national sample of a different society. The observations of Hammond and Centers differ slightly in that the latter found slightly but significantly ($P < 0{\cdot}05$) more men who placed themselves in the working class. This difference may derive from differences in method, or it may be a true cultural difference. At all events, it is trivial. It appears, again, that the 127 men interviewed can be taken, in this respect, as a reasonably representative sample. On the other hand, there is little relation between the worker's opinion of his own social class and his judgements of his work situation. It was found that persons who placed themselves in the working class tended to be satisfied with their tasks; but this tendency was statistically insignificant, and the tendencies for the other three judgements of the work situation were so obviously insignificant that they were clearly not worth pursuing. Even so, the relevance of the worker's own judgement of his social class may be worth considering, as part of a more complex definition of his general view of his place in society.

The second step in this part of the analysis is to consider the worker's general opinions of the three main social classes, whether these three were presented in his own terms (as in most cases) or in the investigators' terms (in a few cases). The summarized opinions are set out in Table 23; and the comparisons between them are interesting. Compared to the middle class, the upper class is disapproved in general (categories 2 to 5, against category 1 : $P < 0{\cdot}001$). Compared to the working class, however, the upper class is merely more vaguely described (categories 2 and 5, against categories 1, 3 and 4: $P < 0{\cdot}05$). Finally, the middle class is better thought of on the whole than is the working class (categories 1 and 5, or approved and neutral ignorance, against categories 2, 3 and 4, or qualified and mixed opinions plus plain disapproval: $P < 0{\cdot}0001$). These are summarized differences which do not depend on exact comparisons among the five separate categories. Even so, there is an obvious tendency for the middle class to be more highly regarded (categories 1 and 5, unqualified approval or neutrality) than is either the upper class or the worker's own class, the working class. This may suggest some preference on the worker's part for promoting himself in the social scale: he may prefer the middle class, however he defines or conceives this, while he can hardly aspire to the upper class (in which no worker placed himself, as was remarked above). On the other hand, it was found that the worker's judgements of these three social classes were entirely unrelated to any of their judgements of their situation at work. As it seems that the worker makes coherent and intelligible judgements of these three social classes, within the limits of his experience, the possibility remains that his comparative judgements of two classes at once may have some relation to his judgements of his situation at work.

TABLE 23

Workers' general opinions of three social classes

The worker's opinion is	Upper class	Middle class	Working class
1. Unqualified approval	34	50	35
2. Qualified approval (good if they respect the rights and position of the lower classes/if they do their duty, try to get on, etc.)	26	16	19
3. Mixed (sympathize with or envy higher classes, but also criticize/approve working class but criticize its lethargy, lack of unity, etc.	15	10	28
4. Unqualified disapproval	21	8	24
5. No opinion: firm assertion that there are no social classes plus plain ignorance	31	43	21
Total	127	127	127

The third step in this part of the analysis is therefore to consider the worker's class preferences, which are easily defined. The five categories of Table 23 can be arranged in a crude rank order from rank 4 (strong approval) to rank 0 (strong disapproval); and the corresponding order of categories is 1, 2, 5, 3, 4. The ranking of category 5 (neutrality, indifference or ignorance) is perhaps disputable, as with any data that are not clearly representable in one 'dimension' only. However, it makes no practical difference to the following analysis whether this category is ranked third (as above) or fourth (after category 3). Once a rank order is established (or alternative rank orders are established), the three pairs of classes can be considered in turn. The finding was that the comparative judgements of upper and middle classes, and of upper and working classes were entirely unrelated to the worker's four judgements of his situation at work. For the comparative judgement of the middle and lower classes, there were interesting but statistically insignificant traces of relations: it was found that persons who clearly preferred the working class to the middle class (i.e. who placed it at least one rank higher, in either scale) tended to be satisfied with their task and with their workmates. Vague as these insignificant findings are, they are intelligible. The worker's opinion of the upper class may be disregarded for uncertainty; but if he prefers his own class to an imaginable hierarchically superior class, it is to be expected that he might judge favourably at least some of the elements in his situation at work. It follows that this particular comparative judgement is also worth considering as a possible part of a more complex definition of the worker's place in society, as he sees it.

The fourth step in the analysis is therefore to consider the worker's classification of persons in ten specified occupations. There have been considered already what the worker is, as he states it, in terms of social classes, and also what he would like to be, in terms of his preference for his own class (the working class) or the next higher class (the middle class). The last consideration is where the worker places ten persons of specified occupational grades who would be placed by the detached observer as belonging to the working class (the stoker, process worker and fitter), the new middle class (the foreman, shop assistant, clerk, executive and production manager), the real middle class (the owner of a small shop), or the upper class (the owner of a large business). The worker's judgements of persons in these occupations may be summarized under five heads. They belong in the lowest of three specified classes (usually called the working class, but excluding the slum class, etc.), or in the middle one of three specified classes, or above the middle one of three; or their position is ambiguous in that it might be either of two possible classes; or, lastly, no placing in terms of social classes is possible because there are no social classes. It was found that there were statistically insignificant traces of a relation between class specifications and the worker's judgement of his situation at work for the first three occupational groups only. For each of these three, the only groups that clearly belonged to the working class as defined by the observer, it appeared that persons who regarded them as working tended possibly (but not significantly) to take a favourable view of their tasks. For the other seven occupational groups there were no systematic tendencies (whether statistically significant or not) that were worthy of notice. It may be concluded that the worker's judgements of his three actual or potential peer groups at work are worth adding to make one judgement. When this was done, a significant relation was found with the worker's judgement of his task only: workers who place all three of the specified occupations (stoker, process worker and fitter) firmly in the working class tend significantly ($P = 0{\cdot}05$) to be those who take a favourable view of their tasks.

The general conclusion is that there are three possible definitions of the worker's social class as he sees it which may by themselves have doubtful (because insignificant) relations with his judgements of his situation at work, but which may possibly be combined to make a single synthetic definition because they all tend in the same direction. The worker has views on the class position of his actual or potential peers, on his class preferences and on his own class position. He is prepared to say, in effect, that they (his actual or potential peers) are in such and such a class, that he prefers this or that class, and that he is a member of a specified class. It is possible, therefore, that these three judgements together may be related to, or determinative of, his judgements of his situation at work. The three judgements are summarized in Table 24. In this table, the judgements are defined as above, except that the

TABLE 24

The worker's judgements of three aspects of social classes

	The worker states that			
	His peers are	His preference is	His class is	Frequency
1.	Middle	Middle	Middle	17
2.	Middle	Middle	Working	17
3.	Middle	Working	Middle	6
4.	Middle	Working	Working	0
5.	Working	Working	Middle	5
6.	Working	Middle	Middle	17
7.	Working	Middle	Working	11
8.	Working	Working	Working	54
			Total	127

last, the person's judgement of his own class, is sharpened. All persons are entered as being working class, in their own view, unless they positively exclude this; and, conversely, only those persons are entered as being middle class who positively state this. Thus the table shows the detail for three simultaneous and dichotomous judgements of the worker's class position, as he reports it indirectly or directly. It is easily inferred from this table that the three judgements are significantly and relatively highly associated. The workers who regard their fellow-workers as members of the working class tend also to prefer the working class, and vice versa; the coefficient of association, (ϕ), = + 0·46. The workers who place their fellows in the working class tend also to place themselves in this class (ϕ = + 0·40). Lastly, the workers who prefer the working class tend also to place themselves in this class ($\phi = + 0{\cdot}30$). Thus it would seem that the synthesis of the worker's judgement of his own place in society may be made intelligibly in the terms of the categories of Table 24. This synthetic judgement, however, is not closely related to the worker's four judgements of his situations at work. It was found that workers whose own placing was clearly and definitely working class (category 8 only of Table 24) tended significantly ($P = 0{\cdot}05$) to be those who were satisfied with their tasks. The other associations were psychologically intelligible though statistically insignificant: satisfaction with pay is slightly associated with the worker's more general preference for the working class (categories 3 to 8 of Table 24), satisfaction with workmates shows the same general tendency as for satisfaction with the task, and satisfaction with the boss is slightly associated with the worker's preference for the middle class (or categories 1 to 4 of Table 24). It is clear that these slight traces of associations between judgements give no ground whatever for inferences about the way in which the worker makes his judgements of his situation at work. What they do give, however, is some ground for the further analysis of the worker's judgements of his place in society. His own present definition of this place is sufficiently detailed in Table 24. It may be, however, that persons who appear in the same category in this table have reached their present conclusions on the basis of substantially different experience: or that some persons have reached different conclusions on apparently the same experience. The general possibility is that the person's present synthetic judgement of his social class may be significantly differentiable with respect to his social or occupational history.

THE WORKER'S OCCUPATIONAL HISTORY

If it is provisionally assumed that the person's occupational history may be relevant, definition is easy. The interview records include both the person's statement of his father's main occupation and of his own previous occupation, with his statement of why he continued in a particular occupation, or at a particular level, or why he did not do so, as the case may be. There is on record, that is to say, both what the person did and why he did it, as he saw the relevant past events when he was asked to consider them in the interview. The relation between the reported fact and the reported reasons given is shown in Table 25. As this table could be set out only with shorthand statements of its categories, it must be read by reference to the following definitions. The thirteen rows show the person's occupational history, as he reports it:

in each row the first letter denotes his father's main occupation, the second his main previous occupation and the last his present occupation. These three occupations are classified thus: I signifies an independent or self-employed person (or a member of the real middle class), N signifies an employed professional or clerical worker (or a member of the new middle class), T signifies a tradesman, and P signifies a process worker or labourer. The ten columns of the table show the person's reported reasons for doing whatever it was that he says he did. These reasons are:

1 He himself decided to improve his place in society, or to stay where his father was, without any substantial help from his father or other adults.

2. He was pushed into a trade because his father, or some other adult in an effective position, thought he would be more secure as a tradesman.

3. He attained the legal and conventional status of a tradesman late in life, by his own efforts as a war-time dilutee (a person who was accepted as a tradesman under emergency conditions, because his experience and skills but not his formal qualifications entitled him to this grading).

4. He left his father's business to become a manual worker because of recurrent quarrels with his father.

5. He did the best he could for himself, without support, because his father was dead or absent, and therefore took up a trade as the next best to an independent or middle-class occupation.

6. He finished as a process worker or labourer through various stages, because he could not be bothered to hold a more responsible job.

7. He finished as a process worker or labourer, through various stages, because he decided that he would thus earn more.

8. He finished as a process worker or labourer because he was impelled to this by his early circumstances (e.g. cost of being apprenticed and lack of openings for apprenticeship to a trade).

9. He became a tradesman by family succession: his father was a tradesman, and neither he nor his father had any other idea than that he should be a tradesman (though not necessarily in his father's particular trade).

10. He became a process worker or labourer by family succession: his father was one or the other of these and neither he nor his father had any other idea than that he should work at the same low occupational level.

It is clear that the relation between the person's occupational history and his reported reasons for his course of conduct are sharp and distinct, though not unique: there may be several (major and minor) reasons for the same course of conduct. It follows, then, that the person's course in his occupational history depends closely, though not solely, on his reasons for the adoption of this course, as he now reports them. His occupational history may therefore be intelligible in the terms of common experience; but, before it can be interpreted in these unscientific terms, it may be as well to show that the workers interviewed reported the same

TABLE 25

Summary of occupational histories

What the worker did	Why he says he did it										
	1	2	3	4	5	6	7	8	9	10	Total
1. PTT	8	7	—	—	—	—	—	—	—	—	15
2. PPT	—	1	4	—	—	—	—	—	—	—	5
3. TTT	2	—	—	—	—	—	—	—	14	—	16
4. ITT	—	1	—	3	3	—	—	—	—	—	7
5. NTT	—	1	—	—	5	—	—	—	—	—	6
6. PPP	2	—	—	—	—	—	—	—	—	15	17
7. PTP	—	—	—	—	—	2	8	—	—	1	11
8. ITP	—	—	—	—	—	—	7	—	—	—	7
9. IPP	—	—	—	2	—	—	8	—	—	—	10
10. NTP	—	—	—	—	—	2	5	—	—	—	7
11. NPP	—	—	—	—	—	2	1	—	—	—	3
12. TTP	—	—	—	—	—	—	8	2	—	—	10
13. TPP	—	—	—	—	—	9	—	4	—	—	13
Total	12	10	4	5	8	15	37	6	14	16	127

kinds of histories as might be expected of manual workers in general. Table 26 summarizes the comparative data, for the historical categories only (that is, the rows only) of Table 25. Besides the present investigation, these data derive from a carefully selected sample of workers (Hammond, in Oeser and Hammond, 1954, p. 239), from two general American samples (Centers, 1948, pp. 198 and 203) that are differentiated by reference to the person who gave the information on occupational history (the fathers or the sons), and lastly from a number of workers who wanted vocational guidance in New York during the pre-war economic depression (Wren, 1942). The categories are condensed from those of Table 25 because the other investigations were concerned only with the son's present occupation (in Center's case, as reported by the father for one son only and as reported by a quite distinct group of persons who also reported their fathers' occupations). In consequence, the middle term in the categories of this table is redundant.

TABLE 26

The worker's occupational history, in five investigations

The worker's history is	Frequency observed in				
	This investigation	Oeser and Hammond	Centers (fathers)	Centers (sons)	Wren
1. PPT *or* PTT	20	7	21	52	11
2. TTT	16	10	16	58	35
3. ITT	7	11	6	27	22 (3 and 4 combined)
4. NTT	6	3	15	46	
5. PPP *or* PTP	28	18	56	92	53
6. ITP *or* IPP	17	7	8	10	79 (6 and 7 combined)
7. NTP *or* NPP	10	4	17	24	
8. TTP *or* TPP	23	6	15	35	102
Total	127	66	154	344	302

It is interesting that the observations from these somewhat diverse sources do not differ greatly. The men interviewed differ from those in Hammond's general sample only in that the latter were significantly ($P = 0{\cdot}02$) more often in category 3 of Table 26, or tradesmen whose fathers were self-employed: a difference which is trivial. Centers' samples include significantly more men in categories 4, 5 and 7 (fathers' report: $P < 0{\cdot}001$) or in categories 2, 3, 4 and 5 (sons' report: $P < 0{\cdot}001$). The two samples, in fact, are not equivalent: the second (from the son's report), which is presumably the more representative, includes significantly more men in categories 1 to 4, or all tradesmen ($P = 0{\cdot}002$). However, some meaning may be attached to the difference that appears in both cases, in categories 4 and 5. It may be that the present investigation did not include a proper fraction of tradesmen whose fathers were white-collar workers, or of unskilled men whose fathers were the same. On the other hand, it may be that the differences reflect general national differences. In any case, although the differences as a whole are significant, they are not marked. The most marked difference ($P = 0{\cdot}0001$) is between the men included in the present investigation and those reported by Wren, who were more often in categories 6 to 8 of Table 26. This is only what one would expect from the circumstances: the men who have come down in the world most seriously are those who seek vocational guidance. It can be concluded, therefore, that the persons interviewed were, in respect of their occupational history, a reasonably representative general group.

The question remains whether their history has any relation to their judgements of their present situation at work. History has been defined in two distinct but strongly associated ways: from the worker's statement of three successive occupational levels and from his description of the critical points, if any, in his movement through these levels. If either definition is related to his judgements of his work situation, this must be because he takes his judgement of his history into his general judgement of his place in society. It is therefore appropriate to consider reduced versions of the two rather elaborate definitions, in which the reductions are made, on the basis of common experience, by reference to the worker's probable judgement of his history. Several such reductions were made for history in the sense of critical points, but are not detailed because none was clearly and obviously related to the worker's judgement of his present situation. Two traces may be noted. A doubtfully significant tendency was found for men who dislike their tasks to be those who became process workers or labourers by the force of various social circumstances (columns 7 and 8 of Table 25). Another doubtfully significant tendency was for these same persons, together with the persons who accepted a drop in social position without any fight (columns 4, 5 and 6 of Table 25), to be satisfied with their bosses as a whole.

For history as defined directly by occupational levels, only one reduction is worth considering, In Table 25, rows 1, 2 and 7 include the men who have achieved and held, or have achieved and failed to hold, a significant rise in occupational position; the persons, presumably, who have some aspirations towards social advancement. Row 3 includes tradesmen by solid succession, about whose aspirations or opinions no inference can be made. Rows 4 and 5 include tradesmen who have dropped in occupational level, and who may or may not be content with their present position. Rows 8 to 11, similarly, include process workers or labourers who have dropped in their occupational level, and who are perhaps not very likely to be content with this fact. Lastly, rows 6, 12 and 13 include process workers and labourers of working-class origin, though some (rows 12 and 13) have dropped in level within the working class.

For this reduction to five main historical classes, the observed relations with the worker's judgements of his work situation are interesting, even though they are of doubtful statistical significance (and, for this reason, are not detailed in tabular form). The workers who are satisfied with their tasks tend to be those who are tradesmen or process workers or labourers by family succession, or those who are now tradesmen but whose fathers were in white-collar occupations or were self-employed. The workers who are satisfied with their pay are all except the tradesmen who have risen from a lower occupational level; and this is the only finding which is statistically acceptable, in conventional terms. The workers who are satisfied with their bosses tend to be the process workers and labourers in general, whatever their history Lastly, the workers who approve their workmates tend to be those whose present occupation, at any of the three levels distinguished, is determined by family succession.

Clearly, these observations cannot be used directly in the interpretation of the worker's judgement of his work situation, because they are of too low a statistical order to be relied upon. On the other hand, all of them are psychologically intelligible, perhaps rather more so than are the equally doubtful, or more doubtful, findings for history defined by the description of its critical points. It may be recalled that similar statistically doubtful but psychological intelligible relations were found between the worker's synthetic judgement of the classes of his society and his judgements of his situation at work. The possibility that was mentioned above can now be taken up: that the effective definition of the worker's place in society, as he sees it, is a compound of his present general view of social structure and his occupational history.

THE WORKER'S PLACE IN SOCIETY

It might be expected, then, that there is some intelligible association between the two elements of the worker's general estimation of his place in society. This was found to depend on which of the alternative definitions of his occupational history was used. It was found that history, as defined by the worker's description of its critical points, had no clearly significant or interpretable relation to the worker's final views on social structure; and this definition of occupational history was therefore discarded. On the other hand, a marked and intelligible relation was found between the worker's history, defined by successive occupational levels, and his final views on social structure. The details are shown in Table 27. This is a condensation of Tables 24 and 25 together, as shown by the numbers of the categories of these tables which are inserted in Table 27. The associations observed are clear and simple. The men who place themselves definitely in the middle class are those who state this explicitly as their own class (categories 1, 3 and 5 of Table 24), together with those who do not make this statement but who both class their workmates in the middle class and prefer the middle to the working class (category 2 of Table 24). These men tend to be those who have worked their way up within the sociological class of manual workers (row

1 of Table 27) and those who have maintained their superior standing in this class by family succession (row 2 in Table 27). The men who would like to belong to the middle class are those who make only tentative statements of their middle class position (categories 4, 6 and 7 of table 24). These men tend to be those who have come down in their occupational level (rows 3 and 4 of Table 27). Lastly, the men who place themselves definitely in the working class on all three counts (category 8 of Table 24) tend to be those who are process workers or labourers of working-class origin (row 5 of Table 27).

TABLE 27

Perceived social structure and reported history

The worker's history is	Table 25 ↓	Table 24 →	The worker's placing of himself is			
			1 Middle class (1, 2, 3, 5)	2 Would like to be middle class (4, 6, 7)	3 Working class (8)	Total
1. PTT, PPT, PTP	(1, 2, 7)		17	3	11	31
2. TTT	(3)		10	3	3	16
3. ITT, NTT	(4, 5)		3	5	5	13
4. ITP, IPP, NTP, NPP	(8–11)		7	11	9	27
5. PPP, TTP, TPP	(6, 12, 13)		8	6	26	40
		Total	45	28	54	127

Besides that these tendencies are psychologically intelligible, they are highly significant statistically. For the 3×3 reduction of Table 27 (combining rows 1 and 2 and rows 3 and 4), the coefficient of contingency for the association between the worker's history and his judgement of his own social class is $+ 0{\cdot}41$ ($P < 0{\cdot}0001$). Evidently, it is the worker's whole occupational history that determines (but only partly determines) his estimation of his own social class. The psychological interpretation may be ventured that the persons who are proud of their superior standing within the working class tend to be those who place themselves definitely in the middle class, while those who once were in the middle class (as an observer would define it) tend to regret that they are no longer so, and therefore to make only a tentative claim for middle class standing; and the men who have the lowest position within the working class tend to accept their position by defining it as being working class without doubt. At all events, a synthetic definition of the worker's place in society can now be attempted, and can be related to his judgements of his work situation as shown in Table 28. This table has been very much condensed so as to avoid the replication of tables for the worker's four judgements of his work situation separately. The table therefore requires detailed explanation. On the left, the table shows the fifteen categories of the worker's place in society that are set out in Table 27; the five kinds of history are each sub-classified by the three kinds of judgement of the worker's own social class. After that, the observed

TABLE 28

Place in society and judgements of the work situation

The worker's place in society is			The incidence of judgements is								The worker is satisfied with
			Task		Pay		Bosses		Mates		
			+	−	+	−	+	−	+	−	
1.	Tradesman, by promotion, possibly unsuccessful	He is middle class	9	8	10	7	5	12	6	11	—
2.		He would like to be middle class	3	—	1	2	—	3	1	2	—
3.		He is working class	6	5	5	6	5	6	8	3	M
4.	Tradesman, by family succession	He is middle class	7	3	9	1	2	8	7	3	T, P, M
5.		He would like to be middle class	2	1	1	2	1	2	2	1	T, P, M
6.		He is working class	3	—	2	1	1	2	2	1	T, P, M
7.	Tradesman, come down in society	He is middle class	2	1	3	—	1	2	—	3	T, P, B
8.		He would like to be middle class	4	1	4	1	3	2	4	1	T, P, B
9.		He is working class	3	2	4	1	1	4	2	3	T, P
10.	Non-tradesman, come down in society	He is middle class	3	4	4	3	4	3	4	3	P, B
11.		He would like to be middle class	6	5	9	2	5	6	4	7	P, B
12.		He is working class	7	2	7	2	3	6	2	7	T, P
13.	Non-tradesman, by family succession	He is middle class	5	3	6	2	6	2	3	5	P, B, M
14.		He would like to be middle class	2	4	4	2	3	3	4	2	P, B, M
15.		He is working class	22	4	19	7	9	17	14	12	T, P, M
		Total	84	43	88	39	49	78	63	64	

incidence among 127 workers of each of these fifteen categories is divided between those who were and those who were not satisfied with each of the four aspects of their work situation in turn. These are the raw data. A summary of the data is shown in the last column of the table. This is derived, for the most part, from the observed statistical tendencies: but it must be made clear that the tendencies noted (for example, in category 3, M signifies that the workers in this category tend to a favourable view of their workmates, but not of their task, pay or bosses (T, P or B as entered elsewhere)) depend also on general intelligibility, or coherence through the whole set of fifteen categories. Where it was necessary, the judgement was made that a category should be classified with another or with some others, because it belonged there on the grounds of common experience or general knowledge, and in spite of the observed statistical tendencies. It follows that the classification of relations between the worker's judgements of his work situation and of his place in society which is shown in the last column of Table 28 is a compromise between general intelligibility and the purely statistical indications. It is this compromise which has to be interpreted; and the interpretation may be referred directly to the worker's judgements of his situation at work, or indirectly to the consequences of his estimations of his place in society. These complementary interpretations may be set out in turn.

(*a*) *Task.* The workers who are satisfied with their present tasks are tradesmen by family succession or tradesmen who have dropped in occupational level, but not persons who have tried to enter a trade, whether successfully or unsuccessfully. The distinction, which is between rows 4 to 9 of Table 28 and rows 1 to 3, holds regardless of the worker's identification with the middle or working class, as he defines these, and therefore depends, presumably, on his acceptance or resentment of his final class position. Further, the workers who are satisfied with their tasks are the process workers or labourers who place themselves in the working class, regardless of their history: and those who are not satisfied are the process workers or labourers who place themselves definitely in the middle class or who would like to belong there. The distinction is between rows 12 and 15 of Table 28 and rows 10, 11, 13 and 14; and this distinction again suggests acceptance or resentment of the final class position, though on somewhat different grounds. The generalization proposed is significant ($P < 0{\cdot}01$), and holds for 79 of 127 workers, or 62 per cent.

(*b*) *Pay.* The workers who are satisfied with their pay are all except the tradesmen who have raised their occupational level or who have unsuccessfully tried to do so (rows 4 to 15 of Table 28, against rows 1 to 3). This distinction would appear to depend only on the worker's social ambitions or lack of ambitions. The generalization is significant (P = 0·025) and holds for 87 of 127 workers, or 68 per cent.

(*c*) *Bosses*. The workers who are satisfied with their bosses in general are persons who have come down in their occupational level, so long as they still regard themselves as members of the middle class or would like to be in this class: while those who are not satisfied with their bosses are the persons who have come down in their occupational level and who now regard themselves as being definitely members of the working class. The distinction here is between rows 7, 8, 10 and 11 of Table 28, against rows 9 and 12. The same distinction appears among process workers and labourers by family succession: if they place themselves in or prefer the middle class (rows 13 and 14 of Table 28), they are satisfied with their bosses while if they place themselves in the working class (row 15) they are not satisfied. Lastly, it appears that the tradesmen who have promoted themselves or who are tradesmen by family succession (rows 1 to 6 of Table 28) are not satisfied with their bosses. The general conclusion is that workers whose position is low in the social scale but who regard themselves as being above this low position are those who accept the authority of management; while those who are low in the social scale and regretfully accept the fact or those who are relatively highly placed among manual workers and who are proud of being where they are, are those who reject the authority of management. This complex generalization is significant (P = 0·017) and holds for 82 or 127 workers or 65 per cent.

(*d*) *Mates*. The workers who get on well with their workmates are those who are of working-class origin and who have not attempted an improvement of their occupational level; while those who do not get on well with their workmates are those of middle-class origin or those of working-class origin who have promoted or attempted to promote themselves in the social scale. The distinction is between rows 4 to 6, 13 to 15 and rows 1 to 3, 7 to 12 of Table 28. The satisfied workers are, again, those who accept their place in society (on the specific grounds stated) and the dissatisfied are those who do not. The generalization proposed is significant (P = 0·025) and holds for 77 of 127 workers, or 60 per cent.

CONCLUSION

It may be suggested, as a general conclusion, that the worker's judgements of his situation at work depend on whether he accepts his relatively low and subordinate place in society, or whether he resents this place, regrets it, or tries to re-define it. The generalizations set out above are somewhat forced and speculative, but all four support this conclusion. However, it seems that the worker's own line of division between acceptance and rejection of his place in society depends on the judgement to which this division refers: that is, the line of division differs for the four separate judgements of the work situation. This variation is to be expected because the worker's place in society, if it can be defined rigorously in his own terms, may be accepted by him or not accepted (in whatever particular terms) in relation to his present purposes, which are various. The dependence of acceptance or rejection (however the latter is defined) on the worker's various purposes may be demonstrated by the consideration of the consequences of the worker's estimation of his place in society; which is complementary to the generalizations that have been proposed so far. For convenience, the complementary analysis is referred only to the case that the worker is not satisfied with the specified elements of his situation at work.

The first point, then, is that workers who have moved up in the social scale and who regard themselves as middle class (definitely, or only by preference: rows 1 and 2 of Table 28) tend not to be satisfied with any one of the four elements of the work situation. These men might be regarded as conscious social climbers, and may be contrasted with those in row 3, who are not so conscious of their potential superiority (as they are still working class) and who therefore still think well of their workmates, though they dislike their tasks, pay and bosses. The solidly superior men (rows 4 to 6 in Table 28), the tradesmen by family succession, dislike their bosses only and, as this holds regardless of their being middle class or working class, the interpretation may be ventured that these are the men who are quite content with their superior position within the working class and who can afford to look down upon the boss.

There is an interesting contrast with the tradesmen of middle class origin. So long as they still regard themselves as middle class (rows 7 and 8), they are content with their tasks and pay and with their bosses, but not with their mates. Apparently they accept their social descent in part but in part ally themselves with the bosses (the men of their former class), against their own class. This does not hold for the tradesmen of middle-class origin who now regard themselves as working class (row 9). These men too are apparently content with their present place in so far as it serves their purposes (the pay for the task) but not with their bosses, because they have repudiated the middle class, nor with their mates because they have not finally repudiated the middle class; thus they still resent being where they acknowledge that they are, even though in some ways it serves their purposes to be there. Exactly the same is true for the process workers and labourers of middle class origin who are now working class, as they see it (row 12).

But, if the men of this last kind of history still think of themselves as middle class (rows 10 and 11), the consequences are again different. These men accept their pay (which, for many, was the chief consideration in their taking up work at a low level: cf. Table 25) and their bosses (the men of their proper class) but, because they are really middle class, they object to their inferior tasks and to the inferior company that they have at work.

The last contrast is among the men who became process workers or labourers by family succession. Those who regard themselves as middle class (rows 13 and 14 of Table 28) accept their position except only for the task (which is inferior anyway): while those who regard themselves as working class (row 15) accept their position except only for the bosses. This latter might be called the conventional working-class view, which is taken also by all the men who became tradesmen by family succession, as noted above.

It has already been observed that these interpretations are forced and speculative. So far as they go beyond what can be strictly inferred from the statistical evidence, they are not to be relied upon; and the consequences of this point, which applies to a lesser extent to interpretations made in other chapters, must be considered in due course (in chapter XI), when all the evidence has been reviewed. None the less, it is interesting that for the findings reviewed, no differences were found by age, length of service or nationality or simply by occupational grade (without history or classification). Of course no differences were likely, as the findings were derived from the whole group of 127 men, regardless of these sub-divisions. The point of interest is that it is possible, though only by elaborate methods, to construct a coherent and intelligible statement of the worker's place in society as he sees it, which appears to hold for a wide variety of men. The conclusions, whether the most rigorous findings or the wider interpretations, cannot be compared with the results of other investigations because no others appear to have been done which included such an elaborate definition of the worker's place in society. Now the point of constructing this definition was in order to elucidate the nature of the worker's judgements of his situation at work. The value of the speculative interpretations of these judgements is just this: if they do amount, altogether, to no more than a provisional hypothesis, at least the hypothesis depends on definite observations instead of general inferences only.

The general conclusion (or provisional hypothesis), then, is that the worker's judgements of his situation at work do indeed depend on his view of his place in society, but not in any simple sense. These judgements do not depend simply on the worker's place in society as an observer would assess this; all the men in question were working men, and it is not surprising that subdivisions within this class were found to be of no direct significance. Neither do the judgements depend simply on the worker's opinion of the whole structure of society as set out in Table 20. In this case, as five distinct opinions were given and as the incidence of these was much the same as in another investigation in the same place, it must be presumed that the worker simply lacked the sophistication necessary to make a judgement of social structure that includes both his own experience and his collection of vague and variously derived stereotypical opinions. Nor do the worker's judgements of his working situation depend on his political opinions or trade union affiliations and sympathies. His political opinions give only an implied and very indirect view of the whole social structure as he sees it. As his opinions are largely cynical, it may be taken that they refer much more to his occasional exercises in compulsory voting than to his immediate personal purposes at work. Similarly, the worker takes a fairly cynical view of trade unions, which is related to his immediate purposes in one particular sense only, the possibility of the union's getting him better working conditions or more pay.

It may be concluded that many workers do not see the class war in conventional Marxist terms, if indeed they see it at all. There are those, although they are only a few, who see themselves as workers oppressed by the capitalist class; but even these few apparently do not take their view into the judgements of the work situation. The interpretation of these judgements in terms of class conflict may be true for some persons: as the analysis carried out referred only to comparisons among working men, this possibility cannot be excluded. But, among working men, it is clear that work is judged in relation to the individual's complex and various ideas of society and not in relation to a sociological abstraction proposed by an observer. The worker's direct, though complexly reconstructed, opinion of his present social class (middle or working) is relevant. So is his occupational history, or the detail of the main stages by which he arrived at his present position, with further detail of the critical points in his progress as a minor qualification. But his history is relevant chiefly in relation to his opinion of his present social class. In very general terms, it could be said that some workers are content with where they have got to in society and some are not. But there is no simple relation between contentment and the worker's estimation of his own class or his history; contentment depends on the complex relation between history and present position, as seen by the worker. Finally, the worker's judgements of his work depend on the same complex relation, with the further complication that the particular dependence varies from one judgement to another among the four that were considered.

It can be said, then, that the provisional hypothesis is a hypothesis about direct reconstructions of the worker's judgements of his work. The general hypothesis is that the worker judges the four aspects of his work situation, variously, in relation to his prior, general and continuing judgement (which is itself made on various grounds) that he has or has not attained an appropriate place in society. But, even if the provisional hypothesis is elevated to the rank of a set of firm conclusions, it leaves much to be explained. The relations of this set, whether of hypotheses or of conclusions, to the worker's judgements of his work in terms of the work situation itself must be considered in due course (in chapter XI). Further, the worker's judgement of his place in society is only one reflection of his experience in general, and it may be that other aspects of this experience outside work are also relevant to his judgements of work. This possibility will be considered in chapters VIII and IX.

CHAPTER VIII

THE WORKER'S PAST AND FUTURE

It is clear that the worker's judgement of his place in society only partly explains his judgements of his situation at work. The next consideration, then, is the relevance of three distinct judgements that lie within the field of 'personality', as it would usually be described.

Firstly, the worker may make a general judgement of the persons with whom he associates. He may incline to the general view that other persons are friendly and can be trusted, or to the view that they must be mistrusted because they are inimical. If he does make a judgement of this sort, it may apply indiscriminately to all the persons with whom he associates or, more likely, it may depend on the kind of association; so that separate judgements would have to be specified for the worker's distinct personal engagements at work, in the home, and so on. Whatever the case may be, there is no direct evidence, because the direct elucidation of possible judgements of persons in general or particular was beyond the scope of the investigation. The worker's statements of how he gets on with his bosses and his workmates have been considered, and his statement of how he gets on at home will be considered in due course. All three of these are specific statements; he does or does not get on with the persons in question in various particular ways. These specific judgements may derive from more general judgements of kinds of personal relations, or of persons in general; and the more general judgements (if they are made), may derive in turn from the worker's experience. It is therefore necessary to go back to his account of his upbringing in order to attempt a reconstruction of his present general judgements of persons; for his early social learning presumably has some relation to his present perception of his situation. The reconstruction is therefore bound to be speculative but, as direct evidence is lacking, it must be attempted.

Secondly, the worker's general judgement of his possible future, in relation to his history, may be relevant. He may be able to state the general or specific objects that he might attain before he retires from work or dies. He may have views on the disasters that might overtake him in the future; and, in either case, his judgement of his possible future may be related to his present evaluation of his past. These possibilities can be considered only by reference to the available information. The worker's statement of his projects or plans for the next few years is relevant. His statement of what he will do when he retires from work, and of what he would do if he had not to work for a living, are also relevant. His evaluations of the best and worst things that might occur in the future must be taken into account because these, in particular, may give direct evidence of his estimation of his possible future. Lastly, it may be necessary to consider the worker's evaluation of his past, in terms of its best and worst features, events or processes. The various kinds of information (points 53 to 59 in the interview check list in the Appendix) may give, at the one extreme, the worker's considered judgement of the whole course of his life and of his prospects; at the other extreme, it may be found that he has only a vague and inarticulate estimation of his possible future.

The third and last aspect of 'personality' that must be considered is the worker's hierarchy of personal values. Whether or not he can make a clear statement on which of his activities matters most to him, the inference from his detailed description of his various engagements should be possible. Presumably what matters most to him, or what he should do above all else, as he judges it, (even if only inarticulately), is the engagement into which he puts his greatest effort or the greatest part of his activity. The relevant information, then, is the worker's statement of what he does, in general, at home; his general methods of, and ideas about, the conduct of his family affairs. This first and most obvious kind of information is available for the married men, but can be backed up by the worker's statement of his whole range of detailed activities outside work. This accessory information is available for all men and, for the married men, it may be expected that it is related to the description of behaviour in the family. It may be possible, therefore, to infer a set of personal values from the combined information on what the worker does outside his work: for, presumably, what he spends most of his time or effort in doing is what matters most to him.

The three possibilities, then, are general judgements of the worker's present situation, in specific and distinct respects. The judgement of persons in general, or of persons in one kind of engagement in particular, if it can be inferred from the worker's account of his upbringing, is perhaps the most specific. The judgement of the possible future, if it can be inferred from the worker's estimation of the hypothetical situations presented in the interview, is likely to be less specific because of its derivation from relatively inarticulate statements. Lastly, the judgement of relative personal values is likely to be the least specific, as it can only be remotely inferred from the worker's detailed account of his activities. Any of these purely hypothetical judgements (or reconstructions of recurrent covert behaviour) could be regarded as 'dimensions of personality'. That these three, and these only, are proposed for consideration depends on the limitations of the information available (as detailed below); but these limitations depend, in turn, on the design and the underlying assumptions of the investigation.

The investigation was not intended to explore 'personality structure' in the usual abstract sense, so that the use of personality tests (inventories and questionnaires) of the abstract kind would have been irrelevant even if it had been practically possible in the circumstances, which may be doubted. The chief aim of the investigation was to reconstruct the person's covert behaviour; the behaviour which occurs between perception of the situation and action, and which links perception and action or directly leads to action. The three kinds of covert behaviour outlined above are therefore relevant to this aim. Indeed, together with the worker's judgement of his place in society (chapter VII), they include most of the general and recurrent covert behaviour (if there is any such behaviour) that would lead on to action in the main engagements of daily life. The whole range of reconstructions would include, in principle, where the worker belongs in his whole society, how he regards other persons, where he thinks he may get to, and what he chiefly values in his present engagements. The only substantial addition that could be made to this range would be the worker's judgement of how he gets on with handling objects (tools and materials). Important as this might be for some purposes, it could hardly be examined economically as a possible general judgement; and the particular judgements that are made in relation to the worker's task have already been considered (chapter IV).

It can be taken, then, that the kinds of covert behaviour which might be inferred from the analysis of the data are the three kinds that are most relevant to the purposes of the investigation. None the less, two related qualifications must be stated about these proposed reconstructions. As they are outlined above, all of these are very broad and general. The worker's possible judgement of other persons in general, if it can be inferred at all, would presumably derive from separate judgements of persons in particular engagements: work, family, friends and so on. The assessment of personal expectations in general should derive, similarly, from the definition of particular projects, aspirations or goals at work, in the home, or within the worker's purely personal (self-directed) activities. The definition of the worker's general values, again, should derive from an inferred rank order of the apparent importance to him of a number of distinct engagements: with his society as a whole, with his work situation (variously), with his wife, his children, and his friends. The reconstruction of the worker's covert behaviour is not referred to this level of detail simply because the information available is not sufficient; as will appear in the detailed account, below, only the broadest generalizations are possible.

A more important qualification is that the proposed inference of three kinds of general covert behaviour would depend much more on inferences drawn from the worker's report of his behaviour (overt or covert) than on direct questions about his aims, expectations or values. It must be allowed that the investigation made too little use of the hypothetical question as defined elsewhere by the author (1957, pp. 116–20), both because this procedure was regarded with somewhat excessive scepticism and particularly because its extended use would have required a much longer interview. The consequence is that the inference of the worker's general covert behaviour, apart from his estimation of his place in society, cannot be derived from the full range of information that would have been desirable, but is derived from a reasonable working compromise. The simplest case is the inference of the worker's expectations or goals, or judgement of the course of his life. This depends partly on his statement of his actual plans or projects for the near future, including his retirement where that was close at hand; and this may be taken to be present recurrent covert behaviour that the person can state openly and clearly if he is asked to do so. The inference also depends, however, on his judgement of hypothetical situations, or situations that he can imagine but has not considered previously: what he might do when he retires (if that is far ahead), what he would do if he had no need to work at all, and what would be the best and worst things that could happen to him in the future. Lastly, the inference may depend, too, on the person's report of the best and worst things that have happened to him; or on his present active revaluation of his memories of the past. In this case, the whole range of information is available.

For the person's central values, the case is different. The information includes the person's report of his behaviour at home and elsewhere outside work, in considerable detail. This information is necessary, for the inference of what matters most to the person must be referred, eventually, to what he does or, as an approximation, to what he says he does. It would have been appropriate to back up this information by asking hypothetical questions about the relative importance of the person's various engagements, or even by using an elaborate questionnaire in which these were contrasted (e.g. the person is required to decide between a better job and his family engagements). The latter was tried in 1951 (factory F only) and was found to be useless in practice both because it required too much time and because its results were inconclusive. The former possibility, the use of direct questions about the relative importance of the person's various engagements, was not tried because it was thought that questions of this sort would be beyond the intellectual sophistication of the men concerned. The inference of a hierarchy of personal values, if there is one, can therefore be made only from the person's report of what he does; but, as this is a detailed report, the inference is reasonable even if it does depend only on indirect information.

The inference of the person's general judgement of other persons is different again, and is the most doubtful of the three inferences suggested. The information from which this might be derived is only the person's statement of how he was brought up. This statement was necessary. Even if it signifies the person's present evaluation of his earlier years, rather than a description of these years as they would have been seen by a detached observer, it gives some information on the person's history. Information of this kind is generally relevant. The cause of the person's present behaviour is his intentional evaluation of his whole situation: but the cause of this evaluation lies in his history. If there is any possibility of taking back the account of systematic causes to include an account of their historical derivation (cf. the author's discussion of this range, 1957, p. 180), the person's history must be taken into account, even if this is known only from his present report, which is the only likely case (on the consequent difficulties, cf. the author's discussion, 1957, p. 158). The person's occupational history has already been taken into account (in chapter VII) and his personal history, in the sense of the best and worst things that happened to him will be considered in this chapter. It is therefore equally appropriate to consider his upbringing as he presently reports it. But it would have been even more appropriate, if not necessary, to determine the person's judgements of other persons by asking hypothetical questions about specific personal relations. In the light of his continuing experience over some years, the worker might be able to make a general judgement or a set of general judgements of the persons with whom he associates. He might be able to say that he has generally found his workmates to be friendly, trustworthy or reliable, without reference to his particular present workmates. He might be able, again, to make equivalent statements about his experience of other personal relations, with his wife, his children, his relatives and his friends. The hypothetical questions appropriate to these important topics should certainly have been asked in principle, but could not be asked in practice. In any case, these questions would have required far more time than was allowable within the limits of the investigation.

In the more important cases (that is, in all cases except the person's general view of his workmates) it would have been necessary to ask questions that were totally unacceptable within the framework of a brief biographical interview. The only possibility, therefore, is that the worker's judgement of other persons, in general or in particular, may possibly be inferred from his account of his upbringing. This inference must depend on the assumption that there is some degree of simple repetition in behaviour, from what the person learnt in his earlier years to what he now does in at least his situation at work, and how he evaluates that situation. The inference may be checked to some degree by the possibility of interpreting present behaviour in terms of past reported behaviour, but it is none the less the most remotely derived and the most speculative of the three inferences that might be made.

The present discussion is confined to the worker's past and future, for practical reasons. The worker's report of his occupational history has been considered (in chapter VII) and needs no further consideration. His upbringing, as he reports it, and his present evaluation of his personal experience in general (the best and worst things that happened to him), are both relevant. So is his report of his plans, projects or vague and general hopes and fears for the future. His account of his present engagements is equally relevant in principle, but is left for separate treatment (in chapter IX) because it depends on more detail than can be reasonably included in one chapter. The past and future, then, and the future in relation to the past, may now be considered on the basis of the various information that was gathered; the consideration of the worker's present engagements (other than work) must be deferred.

THE WORKER'S UPBRINGING

The most remote and speculative reconstruction of the worker's covert behaviour may be considered first. The relevant information (from factories A to E only) is the person's statement of how he was brought up in general, and how he got on with his parents and his siblings in particular. The point may be repeated that this is only the person's present evaluation of his situation as a child. There is no possibility of checking this information against the views of a hypothetical observer, but there is no reason, either, to regard the present statements as grossly distorted by re-valuations made in the light of experience.

At all events, the statements made can easily be classified, and the final classification is shown in Table 29. The categories of this table, which are stated in shorthand form, require explanation.

1. The worker was brought up in the usual way, by both his mother and his father, whether or not he had any siblings. The negative counterparts of this category are various, for the minority (25 of 127) of persons who did not grow up in a normal family. The person was brought up by his mother alone because his father was chronically drunk and therefore ineffective, or absent, by his own decision, or certified insane, or dead; or, the person was brought up in a mixed family by a step-mother or step-father, when he was old enough to notice the difference; or, lastly, he was brought up solely by distant relatives (aunts, uncles, grandparents) or in an orphanage. The distinction made is between the normal family and all other situations.

2. The worker was on generally good terms with whoever brought him up: his parents, his mother, step-parents or guardians. The statements made range from: 'Can you imagine a happy family? That was it', to the conventional: 'Got on O.K.' All in this range are distinguished from all other statements, from that he got on only fairly well (with specified difficulties) to the extreme negative statement that he was always in strife, or miserable.

3. Whether the worker's relations with his father and mother were good or bad (as defined in element 2), they were the same with both parents. These cases are distinguished from those in which the worker stated that he got on better with one parent than with the other and, of course, from the cases in which both parents were not present (as defined in element 1).

4. The general methods of the persons who brought up the worker (whether his parents, his mother only, other relatives or guardians) were described by him as being not strict or strict. The distinction here is between the persons who were described as easy-going or who let the boy alone to do much as he pleased and those who were strict or particular (so called) or who insisted that the boy should do as he was told.

5. The particular methods of the persons who brought up the worker included no specific rules, according to his statement, or did include rules. The negative case is that there were stated rules about where and with whom he was to play, what time he was to come in, what his household duties were, what he should do on Sundays, how he should earn his pocket money, and so on. This is distinguished from all other cases in which no such stated rules were reported by the worker.

These five elements in the worker's upbringing were found to be significantly associated, as shown in Table 29: of the ten coefficients of association (ϕ), all are statistically significant ($P < 0{\cdot}05$ to $< 0{\cdot}001$) except only that for the association between elements 3 and 4. The associations observed are psychologically obvious for the most part. It is in the normal family (element 1) that the boy is likelier to grow up happily and to be on the same terms with both his parents (elements 2 and 3). The normal family, however, tends to be more strict in general than is the broken or substitute family, but does not tend to have detailed sets of rules (elements 4 and 5). These findings are somewhat ambiguous, but both are of borderline significance only (P = 0·044). Secondly, the men whose upbringing was happy (element 2) tend to be those whose relations

TABLE 29

Elements in the worker's reported upbringing

The worker reports that	Associations between elements				
	1	2	3	4	5
1. He was brought up by both parents	—	0·32	0·69	−0·20	0·20
2. He was on generally good terms with whoever brought him up	—	—	0·69	0·43	0·25
3. His relations with both parents were the same	—	—	—	0·15	0·23
4. His parents or guardians were not strict, in general	—	—	—	—	0·26
5. His parents or guardians laid down no set of particular rules	—	—	—	—	—
Frequency in the above categories	102	85	84	66	71
Frequency in the negative counterparts of these	25	42	43	61	56

with both parents were the same (element 3). This association is largely given in the definition of the elements, but there were some persons who reported a happy upbringing in a broken home and some who reported an unhappy upbringing in that they were on bad terms with both parents. The other associations of element 2 are obvious: a happy upbringing is one which is not strict in general and which does not depend on exactly stated sets of rules. The worker's reported relations with his parents (element 3) are not associated with the general strictness of his upbringing but are somewhat associated with the absence of specific rules (elements 4 and 5). Lastly, the men who were not brought up strictly tend also to be those on whom no specific rules were imposed (element 4 with element 5).

This, then, is the worker's upbringing as it is inferred from his present statement. The obvious next step is to reduce the five elements to a single general statement, by making a scale which depends on the observed associations. The case for adding the first three elements is clear: the person was brought up by both parents, was on good terms with them and was on the same terms with both. For these elements the possible scale points range from 3 to 0. The case for including elements 4 and 5 is not so clear, as their associations with the other three are of a low order at the best, or insignificant (3 and 4) or negative (1 and 4). However, an alternative scale of all five elements can be made, and for this the possible range of points is from 5 to 0. Both scales were compared with the worker's four judgements of his situation at work. It was found that only one of these, his judgement of his workmates, was related to either scale score: the men who thought well of their workmates tended significantly ($P < 0{\cdot}02$) to be those who stood high on either scale, though the relation was somewhat clearer for the short scale of the first three elements of reports of upbringing.

Why this one judgement only is related to the worker's report of his upbringing is a question that may be deferred until the more important question is considered, whether the summary specification of reported upbringing by scale scores is the most efficient form. The obvious alternative form is a list of the observed combinations of elements, as set out in Table 30. As there are five elements each of which is classified dichotomously, the number of possible combinations is thirty-two, but only ten combinations were reported. These ten are shown in Table 30, with brief verbal labels supplemented by a statement of their derivation from the positive categories of Table 29, and the incidence of favourable or unfavourable judgements of the worker's mates is shown for each combination, or type of upbringing. It is the distribution of these judgements that matters. The simplest and most obvious distinction that can be drawn in Table 30 is between the first four categories and the other six: and this distinction has already been made by reference to the scale scores for the first three elements (3, against 2 or less: which scores can be read off directly in the rows of Table 30). The men who take a favourable view of their workmates tend to be those who grew up in a normal family and who were on good terms with their parents and whose relations with both parents were the same; and the men who take a poor view of their workmates tend to be those in whose upbringing any one or more of these elements was lacking. This general finding is fairly obvious, but may not be sufficient. The next two categories (5 and 6) of Table 30 include men who apparently tend to think well of their workmates, although their report of their upbringing stands low on either of the scales outlined above. If there are statistical grounds for considering these two categories by themselves, then their possible psychological significance must also be considered.

TABLE 30

Types of upbringing and judgements of workmates

The worker's upbringing was (positive categories of Table 29 are noted)	The worker's judgement of his mates is		
	Favourable	Unfavourable	Total
1. Normal, happy, same with both parents, not strict, no rules (1, 2, 3, 4, 5)	20	16	36
2. Normal, happy, same with both parents, not strict, but rules (1, 2, 3, 4)	9	3	12
3. Normal, happy, same with both parents, strict, but no rules (1, 2, 3, 5)	5	5	10
4. Normal, happy, same with both parents, but strict and rules (1, 2, 3)	11	7	18
5. Broken, happy, one or both parents not present, not strict, no rules (2, 4, 5)	6	3	9
6. Broken, not happy, one or both parents not present, not strict, but rules (4)	6	3	9
7. Broken, not happy, one or both parents not present, strict, with rules (no positive categories of Table 29)	3	4	7
8. Normal, not happy, relations with parents differed, strict, with rules (1)	—	10	10
9. Normal, not happy, relations with parents differed, strict, but no rules (1, 5)	1	7	8
10. Normal, not happy, same with both parents, strict, but no rules (1, 3, 5)	2	6	8
Total	63	64	127

On the statistical point, the evidence is clear. The doubtful categories (5 and 6) do not differ from all the other categories together in their relation with the worker's judgement of his workmates ($P = 0{\cdot}19$). But they differ even less from the categories of high scale value (1 to 4: $P = 0{\cdot}75$); and they differ markedly from the other categories of low scale value (7 to 10: $P < 0{\cdot}01$). There is therefore a clear case for considering the possible psychological significance of these categories, as types or patterns of upbringing, regardless of their summed scale values. For both, reference to the interview records easily clarifies the position. The men in category 5 were brought up by their mothers only, effectively, because their fathers were chronic drunkards; and these men, some of whom had to carry quite a part of the family's burdens from an early age, were all on very good and exceptionally close terms with their mothers. It is not unreasonable to suppose that their social learning in emergency, or under pressure, was at least as effective as the learning of the men who grew up happily in well balanced normal families (categories 1 to 4). A similar case may be made for the men in category 6 of Table 30. These men were brought up to a large extent, as they report it, by their elder siblings because their fathers were dead, or absent by desertion, or ineffective in a large family; and the best that their mothers could do to cope with these unequal situations was to delegate a large part of their responsibilities to the elder children. In consequence, the men in this category were not happy in their early years because the whole family was miserably poor, and were brought up under the specific rules that were required by the situation. None the less, it is reasonable to suppose that their social learning, under the considerable pressure of their circumstances, was again as effective as the learning of the men who grew up under no such pressure.

These cases may be contrasted with the cases of the men in the last four categories of Table 30, for which reference to the interview records is again relevant. The simplest is category 7. The men in this category were brought up strictly and with specific rules, as were the men in category 4. Although there is no significant difference between these categories ($P = 0{\cdot}69$), both are perhaps better considered in the context of the other categories, or patterns of upbringing, than in isolation. The psychological speculation is therefore put forward that a strict and regulated upbringing may be acceptable if it is in a normal home in which the parents do not differ (category 4); if it is in a substitute home, with a step-parent, with distant relations or with institutional guardians (category 7), it is too much to accept, and the boy is not happy. The case with category 8 is simpler. This does differ significantly from category 4 ($P < 0{\cdot}01$), and it is clear from the records that it does so on simple grounds. The men in category 8 were brought up by heavy-handed autocratic fathers (to the extent that in some cases the mother's place was merely that of a domestic servant), and had to split their loyalty painfully between the parents; in consequence, they were not happy. Somewhat the same holds for category 9, although this does not differ significantly from its logical counterpart, category 3. For the men in category 9, it was the mother who was on their backs in general, but not specifically: the father was their close mate. The consequence, again, of a painful division of loyalty was unhappiness. Category 10, again, does not differ significantly from its logical counterpart, category 3 ($P = 0{\cdot}56$); but, in this case, the distinction in the interview records is particularly clear. For the men in category 3, the parents' strictness was simply a means for bringing up the boy decently, in accordance with the parents' principles. For the men in category 10, the parents' strictness was an end in itself: their aim was at the least to hold the boy at arm's length so as not to be bothered with him and at the most to suppress him by the use of whatever kinds of assault (as detailed in the records) might be found most effective. This is the clearest case in which the pattern of upbringing has a psychological significance which goes beyond what is given in the mere sum of scaled elements. None the less, as in other cases, the distinction drawn is speculative. The general distinctions, then, are clear even if they are also speculative. If the boy is brought up in a normal happy home (categories 1 to 4), he is likely to grow up to be a friendly man. If he grows up in a home in which he has to divide his loyalty or in which he is merely given his orders and made to mind his place (categories 7 to 10), he is likely to become an unfriendly man. If he grows up in a broken home, he may none the less become a friendly man, under either of two conditions (categories 5 and 6). If the mother and children unite closely to cope with the father's total incapacity or absence, the boy may actually have a happy childhood. Or if the pressure of poverty, and of the detailed restrictions which follow from this, are so great that the boy is not happy, he may still become a friendly man because he has learnt the hard way how to cope with emergencies.

For the general division of Table 30 (between categories 6 and 7) there are no differences in the relation observed by occupational grade, age, length of service, or nationality. No comparison with other findings is possible because this, again, is a topic which appears not to have been examined in detail previously.

The relation observed, if it is a plausible synthetic reconstruction of the whole set of possible relations between patterns of upbringing and present judgements of the work situation, is observed for one of these judgements only, that of the worker's mates. The question must therefore be considered, whether it is reasonable that there should be a relation for this judgement only and not for the other three. The ground of the relation remains to be considered but, whatever it may be, the distinction is in one respect obvious. The worker's report of how he gets on with his mates is a judgement of a personal association which may or may not be close but which is continuous. His judgements of his task and pay are not judgements of personal relations at all, but of object and instrumental relations respectively. If there is any reason to expect an association between upbringing, as reported, and judgements of the present work situation, the expected associations would be for personal relations and not for relations of other kinds. This is reasonable, at the least because the person's upbringing in relation to objects (i.e. his learning of motor skills) was not enquired into, but more generally because his upbringing (i.e. his development in the home) is concerned much more with immediate social learning than with learning about objects or general ideas. It is possible, of course, that object-learning in a special sense, such as anal fixation due to toilet training, may have consequences in judgements of the present task; but this sort of speculation (on which see Jones, 1920, ch. XL) is far beyond the limits of the present investigation.

For the person's judgement of his bosses, the case is not so simple. As an overall judgement, this often does not refer to immediate personal relations. The worker's immediate boss may be little known and his higher bosses may be entirely unknown as persons, and are well known in no case. In any case, the worker's relation with his boss is official firstly and personal only thereafter. None the less, some association might be expected between the worker's report of his upbringing and his judgement of his immediate boss, which is the more personal of the two elements compounded together in the overall judgement. The association observed is not of a high statistical order ($P = 0{\cdot}04$) even for a two-fold rearrangement of the categories of Table 30, but it is interpretable. It is found that men in categories 2, 3, 4 and 10 tend to think well of their immediate bosses and men in the other categories tend to a poor opinion. The distinction is two-fold. Firstly, between men who were brought up in normal homes and who were on the same terms with both their parents, though not necessarily happy (category 10), and the men from broken homes or whose relations with their parents differed (categories 5, 6, 7, 8 and 9). Secondly, among the men in the former of these two broad groups, it is only those whose upbringing was strict or regulated or both (categories 2, 3, 4 and 10) who tend to like their immediate boss: those whose upbringing was neither strict nor regulated (category 1) tend to dislike the boss. A possible speculation therefore is that the immediate boss who is liked may be regarded as an acceptably strict father. In view of the low order of the finding, this is pushing speculation to the limit. But it is not surprising that no clearer statement can be made, as the persons who were judged as immediate bosses ranged from leading hands to production manager and, even in the usual case of the judgement's being referred to the foreman, it was clear that there was considerable variety in the degree of personal contact on which it was based. In any case, this possible finding offers no explanation of the worker's judgement of his bosses as a whole. It may be concluded that this judgement too is not of a direct personal relation which is at all comparable to the worker's judgement of his mates.

The general question must now be considered, what is the

nature of the relation between the worker's upbringing, as he reports it, and his judgement of his workmates. What has been suggested so far is a historical cause of this judgement. If the worker was brought up normally and happily, or if he learnt to adjust himself to continuing emergencies in the home, then he learnt to be friendly; while if he was generally unhappy in his childhood, he learnt to be unfriendly. But this possible link cannot be direct. In fact, stating it in these terms is a departure from all the argument of chapters IV to VII, which was concerned only with the elucidation of the present or systematic causes of the worker's judgements. His judgement of his task, for example, was considered, in chapter VII, in relation to his prior but present and continuing general judgement of his place in society. In chapter IV, the worker's judgement of his task was related even more directly to the elements of the task itself; that is, the judgement was derived immediately from the compounding of separate judgements of aspects of the task.

The distinction between historical and systematic causes of behaviour may be difficult to make in some cases (cf. the author's discussion of this, 1957, pp. 154–9). In the present case, however, there can be no doubt that the worker's upbringing (if his present report is taken literally, as history) is a remote historical cause of his present behaviour and can be linked with this behaviour only through a series of intervening transformations. The only conceivable present or systematic cause of the worker's judgement of his mates is some kind of present general judgement of persons, or some set of present general expectations about persons, that leads on to the specific judgement of the men with whom the person works. Whether this hypothetical construct is called a judgement or a set of expectations, it must be general in some sense, as it cannot be a simple repetition of judgements made in earlier years. In his early years, the worker may have already formulated the judgements that he now reports, but these were then judgements of his parents and his siblings, whereas his present judgement is of men whose relation to him is quite different from his previous relations with members of his family. The hypothetical reconstruction must therefore extend, as it develops in time, to include at the least the worker's present relations with his mates, and possibly other present personal relations too. It may be suggested that a fairly direct transfer of early learning to present judgements is likely, as it is presumably in the happy family that the person gets on well with his siblings. The analysis reported above, indeed, depends on relations with siblings as well as with parents; but the former, unfortunately, were reported in considerably less detail than the latter. Where sufficient detail is given, it is clear that the two relations were usually of the same nature; and specially close relations with siblings are reported by some men who were brought up by the mother only or, in the extreme case, by elder siblings only or mainly. Information on the number of other children in the home (including unrelated children in a few foster-homes) is complete, and may be used to support the suggestion made. Of 127 men, 105 were brought up with 2 or more siblings (or other children), 14 with one other and 8 only were brought up alone. There is no relation between the number of siblings and the general relations reported in the home: what matters, apparently, is not the formal possibility of friendship with siblings, but the psychological possibility. The suggestion made may therefore be taken as plausible; but, even if it could be established, it would not demonstrate a direct transfer. The most that one could say would be that sibling relations are more closely parallel to relations with workmates than are relations with parents.

On the other hand, the reconstruction may not be a simple generalization of early experience, over whatever area. If the information is taken as literal history, the reconstruction must be at least this; but all that one has for certain is a present report of past events, which report is likely to be to some extent a present revaluation of these events in the light of the worker's intervening experience of various personal relations. Further, the reported revaluations of early experience range widely, from a few made almost in the person's contemporary situation (by adolescent apprentices) to a few, at the other extreme, made over almost the distance of two generations. For the elements in the worker's upbringing, except only for his report of by whom he was brought up, some revaluation may be likely in proportion to his present distance from the events which he ostensibly reports. For the family men, whose present experience reflects their past from a different point of view, considerable revaluation is likely; and the family men were the large majority (92 of 127). If past experience reflects into judgements of the present continuing situation, experience in this situation may equally reflect into judgements of the past. These considerations reinforce the possibility that the worker's report of his upbringing may be taken as a report of his present general judgement of, or general expectations of, persons in various relations. This certainly includes his relations with his fellow workers, and may include also other personal relations (as discussed in chapter IX). The former alone, however, is the likelier finding, as the worker's mates are the only persons who are related to him more or less as persons in general, in almost the same incidental and unexacting way as are the persons with whom he has occasional routine dealings outside work (e.g. tram-conductors, shopkeepers). A finding that would include both relations of this kind and the close personal relations that he has entered deliberately is perhaps unlikely.

The final generalization, then, is a remote and speculative reconstruction of the worker's covert behaviour, and it is an approximation only: while it is highly significant ($P < 0{\cdot}0001$), it holds for only 83 of 127 workers, or 65 per cent. However, it can be simply stated. On the basis of both his early experience and his later personal relations, the worker concludes that other persons are generally friendly and forthcoming; or that they are reliable in adversity; or that they are unreliable in general. In the first two cases, he goes on to make a favourable judgement of his workmates; in the last, his judgement is unfavourable.

THE WORKER'S PERSONAL EXPECTATIONS

The information (for factories A to E only) from which one might be able to infer the worker's concrete plans or unlikely hopes for his personal future is diverse. The most definite is the worker's statement of his plans or aspirations for the next few years. A few stated definite plans to improve their present position at work, or to go into business on their own. A few more stated plans to improve their position outside work, by buying or building a house or buying a car, and so on; and some had plans of both kinds. Many more had only limited personal projects in mind (such as becoming a good tennis player, or learning languages) or had only vaguely formulated aspirations (such as to win a lottery and go for a trip). Finally, a large minority (39 of 127) could see nothing before them in the next few years except to go on as they were. Even the immediate future, then, was seen in a variety of ways. Several alternative classifications of this information were made, and it was found that none was clearly related to any of the four judgements of the work situation. There was a just significant relation between one of the classifications made and the worker's judgement of his bosses. It appeared that workers who had clearly formulated and realistic plans for their next few years tended to think worse of their bosses than did the other workers. But this finding depended on a rather forced classification of data that were ambiguous in some cases and, as it was only just significant statistically, it was disregarded.

The next step was to consider the worker's statement of what he would do when he retired. This ranged widely, from statements of defined plans for a retirement that was close at hand (including both where to live and what to live on), to statements made by a few adolescents on what they thought they might do in their future grand-children's time. In consequence, these statements were much less definite than those of the worker's immediate plans, which were referred to the same range of time for all persons. In consequence, again, it was found that a minority only (27 of 127, most of whom were older men) had any definite ideas about, let alone plans for, the last years of their lives. A lesser minority (20 of 127, again mostly older men) could think of nothing more than that they would have to keep themselves occupied, or they would go mad. The remaining 80 of 127 specified various fantastic aims (such as going for a trip around the world) or simply stated that they had no idea what they would do. It was found that the statements made were not related at all to the four judgements of the work situation. In view of the diverse bases of these statements, this is not surprising. Throughout the investigation, it was necessary to work with a group of persons ranging widely in age. In this particular case as in others in which age was probably relevant, the subdivision of the range would have left sub-frequencies that were statistically too small to be of use. The third step was therefore to consider information that was on the same hypothetical basis for all persons. If the worker's plans for his next few years and for his retirement both refer to a future which is certain and which can in principle be planned for, his opinion of what he would do if he had no need to work at all refers to a purely hypothetical situation which may, indeed, be hard for him to imagine. It was found that most of the workers saw this hypothetical situation as one in which it would still be necessary for them to work if only (presumably) in order to keep themselves occupied. Of the 127 men interviewed, 57 specified that they would go into business of some sort (landlord, shopkeeper or farmer; that is, as owners, not workmen). Another 13 (or 70 total, of 127) merely said that they would go on working. The minority of 57 had no ideas beyond taking a holiday or having a vaguely specified good time. These results concur generally with those reported by Sharp (in Oeser and Hammond, 1954, p. 229), although both the question put and the persons to whom it was put were somewhat different in the work reported by Sharp. It was again found, however, that the statements made were not related at all to the judgements of the work situation. Again, this was to be expected. The hypothetical possibility that the worker might be able to retire tomorrow is too remote altogether (even in a country which has numerous well supported lotteries) to bear any relation to his judgements of his present actual situation.

The fourth and last step in the analysis, then, was to consider the worker's statements about the best and worst things that could happen to him. The answers to these questions must depend heavily on the imagination, but not in the same sense as with the questions previously considered. For one thing, the person has only to imagine what is possible, within his own terms of reference, and not what is hardly conceivable (such as winning enough money to retire on). Besides this, the person can refer his imaginary construction to whatever range of time may suit him: he is not bound to consider either the next few years of his life, regardless of his age, or the end of his life, again regardless of his age. These questions, then, might be expected to strike a compromise between the most definite statement of his plans (for the next few years in general, or for retirement if that was due in the next few years) and the most vague statement of hypothetical aspirations.

It was found that the answers to both questions could be classified on a simple dichotomy. The best thing that could happen in the future is simply that the person should go on as he is now, stay healthy, or just go on living; or, at the most, that he should be fantastically and inconceivably lucky and win enough money to retire, be idle and enjoy himself. Alternatively, the best thing that could happen is that he should be able to get on with his particular present projects; to get married, to promote his family's welfare, to educate his children, to improve his home, to improve his own occupational prospects, and so on. The incidence of these two main classifications was 77 and 50 respectively of 127; and this supports the suggestion of a compromise between concrete plans and unlimited fantasy, because there were far less than 50 men who had really definite projects for the near future. At all events, the distinction is clear. The majority of the men interviewed see their future simply as a continuation of their present, unless their lot is accidentally and therefore quite unpredictably improved; while the minority see their future as something that they themselves could possibly shape, or hope to shape, even if they have no definite plans for doing so. The worst that could happen is classified equally simply. On the one hand, the person might be struck down by fate before his time, by chronic illness, by being crippled or by an early death. On the other hand, he fears only such strokes of fate as are inevitable for anyone (bereavements, death in due course), or only remotely unlikely disasters (as they were in 1952: war, revolution unemployment); or, lastly, he may fear nothing which can be particularly specified. The distinction here is between the minority (27 of 127) who are apprehensive of their future on biological grounds and the majority (100 of 127) who are not apprehensive in this sense.

It was found that the two dichotomous classifications were associated ($P = 0{\cdot}024$). The tendency is that the men who see their future simply as a continuation of the present are none the less apprehensive of the future: while the men who expect to be able to shape their own future to some extent are not apprehensive. At all events, the two dichotomies together give four views of the future, or general expectations. It was found that this fourfold classification was significantly related ($P < 0{\cdot}0001$) to the worker's judgement of his bosses. The tendency is for the workers who expect nothing of the future either way (their course will continue, without improvement, but also without disaster) to be contented with their bosses: and for all other workers to be less contented. Interesting as this generalization is, it is worth taking a stage further. Besides the questions about the future, parallel questions were also put to the worker about the best and worst events in his past. The information gathered from both of these was classifiable in a simple dichotomy, but it was found that only the latter was related (and that somewhat doubtfully) to the fourfold classification of possibilities in the future. For the past as now evaluated, then, the worst thing that happened to the person is that he suffered serious personal illness (including war injuries), or was involved in general disasters (unemployment or the personal dislocations, deprivations and bereavements consequent on his going to war); or that he suffered serious personal disasters (the extreme case of which was: 'It happened years ago, but it was too horrible to tell'). Alternatively, the worst that happened to the person was bereavements of the inevitable kind (death of grandparents or parents), or trivial blows at the ego (he was shamed before his class, at the age of 12: but it is clear that he now regards this incident with detached contempt); or nothing at all that he can now specify. The distinction here is between the workers who now regard themselves as having had a bad course of events, whether on personal or on general grounds and those who now regard themselves as having had an ordinary run with no disasters beyond those that are likely to overtake everyone in due course.

If this distinction is combined with those drawn above for the four categories of the worker's view of his future, there are eight categories altogether. It was found that this compound classification of general expectations was even more closely related to the worker's judgement of his bosses (but to that judgement only) than was the classification of the four categories of the future as the worker sees it. The details are set out in Table 31, in which the eight categories have been reduced to five only by the inclusion of all the men who hope to be able to shape their future in category 5, regardless of their views on the worst that might happen to them or on the worst that did happen to them. This reduction is made merely for convenience. There are few or no cases in three of the four categories thus condensed into one, and the distinctions among these four are not necessary to the discussion of the results. The titles of the categories are abbreviated versions of the distinctions drawn above. Thus 'stay as he is' (categories 1 to 4) signifies that the worker expects his future simply to continue his present; while 'go on, get on' (category 5) signifies that he hopes to be able to shape his future. In the second part of the title, 'apprehensive' means that the worker sees the possibility of being struck down before his time as the worst thing that could happen to him, and 'not apprehensive' refers to inevitable or unlikely disasters. In the third part of the title, 'personal' means that the worker sees his past as including serious, painful and possibly unusual misfortunes; while 'inevitable' means that if any disasters are reported, they are merely of the kind that must happen to everyone.

In Table 31, the obvious distinction is between category 3 and categories 1, 2, 4 and 5. In the former, a majority of workers think well of their bosses in general; in the latter four categories indifferently a majority have a poor opinion of their bosses. This highly significant distinction ($P = 0{\cdot}0000003$), which holds for 99 of the 127 workers, requires interpretation. It is suggested that the men in category 3 are the men who have finally come to terms

TABLE 31

The worker's expectations and judgements of his bosses

The worker's opinion is			Judgement of bosses			Average age (years)
Of his future		Of past disasters				
The best	The worst		Satisfied	Not	Total	
1. Stay as he is	Apprehensive	Personal	4	14	18	28·9
2. Stay as he is	Apprehensive	Inevitable	—	4	4	38·3
3. Stay as he is	Not apprehensive	Personal	32	11	43	38·1
4. Stay as he is	Not apprehensive	Inevitable	2	9	11	30·9
5. Go on, get on	Apprehensive or not	Personal or inevitable	11	40	51	30·6
		Total	49	78	127	

with themselves and their fate, on the basis of a varied experience. They are not necessarily content with their place in society, but they accept it as a fact that they will never be other than what they are now, whether in social class, in wealth or possessions or in personal attributes; and they have no fear of being struck down before their time. Further, this double judgement is made after the experience of serious misfortunes. These men, then, are those who have learnt the hard way to take things as they come, and who are therefore content to take their bosses as they come. The men in category 1 also accept the fact that they will never be other than what they are now, and they also do this on the ground of experience of serious misfortunes. But they fear being struck down by fate. They have not yet come to terms with themselves, and are therefore apprehensive, not only of the possible disasters that they mention, but also of their bosses in general, in whose power their continuing future lies to some extent. The men in category 4 have come to terms with themselves in the same way as the men in category 3, but have not yet got to the point that they are prepared to accept their bosses, among other natural hazards, as they come. The distinction here presumably depends on their earlier experience. As they see this, nothing ever happened to the 11 men in category 4 beyond what must happen to any man at some time; and they are therefore not disposed to accept their inevitable bosses. This distinction depends on the assumption that the bosses in general are the worker's natural enemies. The assumption is only to some extent supported by the differences in judgements reported in chapter III (satisfaction with the bosses is the lowest, and is highly significantly lower than satisfaction with the task or with pay, but not significantly lower than satisfaction with workmates). However, the assumption may be made, if only speculatively, on the general ground that the worker's bosses in general are not so much persons as agencies of power that have the possibility of disposing of his future. Alternative explanations of the distinctions between categories 4 and 3 would be that the men in the former are merely too stupid to have worked it out for themselves that the bosses too must be accepted as they come; or too intelligent to accept their natural enemies. On these possibilities, however, there is no evidence. The men in category 2 are like those in category 1 in that they clearly have come to terms with themselves, but it is not obvious why this should be so, as they have no experience (as they see it) of serious misfortunes. Whatever the reason may be, the consequence of the fact is that they distrust their bosses in the same way as the men in category 1. Lastly, the men in category 5 are those who have an eye to the possible future that they might shape for themselves, regardless of the worst things that might or did happen to them. Their views on what they might achieve, at the best, go some way beyond the concrete and realistic projects of the smaller minority who had plans for their next few years, and must therefore be regarded as aspirations (though in realistic terms) rather than projects. Even so, the interpretation is simple: the boss, as a person who can to some extent dispose of the worker's future, is on the whole seen as a bar to the worker's aspirations.

These speculative interpretations may be reviewed with regard to the average ages of the men concerned, as shown in the last column of Table 31. The greatest difference between these is over nine years. The particular differences between categories, with regard to age, may not all be statistically significant, but the overall differences are: the correlation ratio, η, is 0·37 for the five categories set out in the table and also for the reduction to three categories (1, 2 and 3, 4 and 5). In categories 2 and 3, then, are the old men who may or may not have come to terms with themselves. The difference may be due to a real difference in experience: serious misfortunes (category 3) may be the appropriate ground for learning the hard way that the future is not to be feared because it can be coped with, while a relatively uneventful course (category 2) may leave the person open to fear how long his luck will hold. Equally, the difference may be due to elements in the person's history that were outside the scope of the enquiry. In either case, the distinction is clear: the old and (in general) more experienced worker is prepared to accept his bosses as they come if and only if he has come to terms with his own experience. In categories 4 and 5 are the younger workers, some of whom expect nothing in the future (4), while others (5) are still interested in getting somewhere. In this case, the difference cannot be due to differences in experience, as category 5 includes both of the main kinds of experience. It must be presumed, therefore, that the difference depends on personal factors outside the scope of the enquiry. But the practical effect is simple: the men in both these categories have not come to terms with themselves (in quite different ways, for probably quite different reasons), and both groups therefore are not disposed to accept their bosses. Lastly, the youngest men (category 1) have not yet accepted the fact that life in general, and the life of a working man in particular, is often rough, and therefore take a poor view of their bosses.

These interpretations, whether by category or by age, are of course forced to the limit of acceptable speculation. None the less, the main distinction is simple enough. The workers who have accepted, or who are resigned to, their personal fate (though not necessarily their place in society) tend to be those who accept their bosses; and the workers who, in whatever way, do not accept their fate tend to be those who do not accept their bosses. It may be as well to make it clear that this main distinction, while it depends to some extent on age (and therefore on experience) does not depend only on this. It is true that older workers (30 and over) tend more to a favourable view of their bosses than do young workers (29 and under); but this finding, which holds for only 71 of 127 men, is of a much lower order ($P < 0{\cdot}02$) than the finding discussed above. The relevance of age is that experience depends on what the person does with it. For the main finding discussed above, then, there are only subsidiary differences by age; and there are no differences by length of service, occupational grade or nationality. No comparison can be made with the findings of other investigations because the main finding, in this case again, appears to be the only one of its kind.

The finding, however, is that the worker's expectations for his personal future are related to his judgement of his bosses in general, and to that judgement only. Besides the question whether the account given above can be taken as a reasonable reconstruction of the worker's actual covert behaviour, there is the question why one highly significant relation should hold, while the other three judgements of the work situation are clearly independent of the worker's expectations. To the first question, no firm answer can be given, in advance of the general argument in chapter XI. It cannot be said that the elaborate reconstruction of the way in which the worker sees his personal future is a reasonable representation of his actual views. On the other hand, it is reasonable to suppose, on the basis of common experience, that the worker does have views on his personal future. Whether or not these views could be stated in the terms proposed above, it seems likely that there would be some workers who think they may go somewhere, some who know that they will go nowhere beyond where they are, and some who are uncertain. The second question can therefore be answered in general terms.

The worker's judgement of his mates may be considered first, and can be disposed of briefly, on the ground that the worker's relation with his mates is the least exacting of all his relations. He may dislike his mates because they are not the sort of persons a man of his position and origin should be working with (cf. chapter VII). But, although he has to work with other men, this is only an incidental and unexacting consequence of his having taken a job in a particular firm. The most that he must do is to co-operate with other men in the performance of the task and the least that he may be able to do, if he wishes, is to ignore his fellow workers completely. Any worker can presumably do this off duty if he wants to, and some who work by themselves can do it at all times. The worker's behaviour with regard to his mates presumably depends to some extent on his general judgement of persons, as inferred from his upbringing, and on his judgement of his place in society: for his judgement of his mates depends significantly on both of these. But whether or not he acknowledges the fact that he will always be as he is now can hardly be relevant to his judgement of his mates. If he does not acknowledge the fact, he may like or dislike his mates, indifferently, on other grounds; and the reverse holds if he does acknowledge the fact that he has no future beyond his present. In this case recognizing the realities of his position does not oblige the worker to get on with his mates, because he can easily shun them if he wishes; and retaining some hope that he can shape his own future in no way commits him to friendly or unfriendly behaviour at work. His relations with his mates are so easily manageable that, on whatever grounds he may dislike them, he can hardly regard them either as a threat to his future security or a bar to his possible advancement in general.

The case with the worker's judgements of his task and pay is the contrary. His task and pay, far from being variable on his initiative, are likely to be fixed within narrow limits at an early stage of his career. The main decision whether or not he is to be a tradesman has to be made at an early age and, with very few exceptions, is then irrevocable for the men who do not take an apprentice training. It is easier for the man who begins as a labourer to become a semi-skilled worker, but is by no means an open possibility: as time goes on, his continued experience of unskilled work only counts against him more and more. Further, the worker can do little to change his conditions by moving from one factory to another. There are differences between factories, but the major relevant distinctions (as set out in chapter IV) are given in the nature of the task. There are clean trades (such as the carpenter's work) and dirty trades (such as most branches of the engineer's work). There are independent labouring tasks and others that are closely supervised; and there are complete semi-skilled tasks and others that include only a fragmentary set of operations. Of course in the two latter cases the worker can try to move from the less to the more preferable kind of task; but the possibility of his doing so is limited, not so much by differences between factories, as by the rareness of the more preferable tasks of either kind. It seems likely, then, that the worker knows from an early age that he is likely to go on doing the same sort of task until he retires. If he accepts the fact that he will never be other than he is in general, he may still like or dislike his task, depending simply on what kind of task it is, as well as on his opinion of his right place in society. If he is either apprehensive of his personal future or determined to get on, he may like or dislike his task, indifferently, on the same grounds. The given fact of the task may be accepted or resented; but, because it is given and not capable of alteration by the worker's own efforts, it can hardly be seen either as a threat to his security or a bar to his advancement. Much the same would hold for the worker's pay. This is a little more open to alteration, in that the worker can seek a factory which pays somewhat more than usual for any particular task or kind of task. But for the most part pay is determined simply by the task, and therefore is probably accepted at an early stage as an unalterable fact, even though it may be disapproved.

The case with the worker's bosses is intermediate between the other two cases. His relations with his bosses are to some extent manageable on his initiative. For one thing, he can move from one factory to another in search of better bosses; and both factory managements and their foremen vary much more than do tasks or rates of pay. There is no contradiction here of the finding discussed in chapter III that dissatisfactions with the task, pay and bosses are first, second and third in rank order of importance as reasons for the worker's changing his job. His prospects, when he decides to change, are different in the three cases. An experienced worker may know that it may be difficult to find a better task, and that it may be only a little easier to find a better paid job (though he can assess this, and this only, before he enters employment). For the bosses, he can only expect that they may be different; whether better or worse, he can determine only by prolonged trial. That he has the possibility of significant change in this respect, then, is quite compatible with the fact that he is not likely to seek a change on this ground alone; it is the possibility that matters. A more important point is that the worker's relations with his immediate boss or bosses may change and develop on the ordinary basis of personal interaction; so that the older workers may be able to come to terms, not only with themselves, but also with their immediate bosses.

On the other hand, the worker's relations with his bosses are also to a considerable extent fixed, and not capable of alteration on his initiative. The immediate boss is not just another person (like the worker's mates) but a person with authority, who has both the power and the duty to control the worker's actions. The higher bosses, if they are known at all as persons (and they are often not), have this function in a higher degree. The relation of subordination is a fact that any worker must accept as a fact, whether or not he resents it. Besides this, the firm's policy may vary in accordance with market conditions. Where this has reflections in personnel policy, and therefore in the behaviour of the foremen, the worker can only accept the change as a fact or resist it; he takes no part in deciding it. The worker's relations with his bosses, then, are manageable on his initiative over a wider range than are his task or his pay; he can interact continuously with persons, where he can only accept as a given fact the complexly derived things he handles or the legally defined pay system under which he operates. On the other hand, his relations with his bosses are much less open to determination on his initiative than are his relations with his mates; in the one case, he can act only as a subordinate whose views may be ignored, in the other he acts as an equal. His relations with his bosses are therefore manageable to some extent only, with effort, and at some risk. It is thus to be expected that he may see his bosses as a threat to his future security or a bar to his projected advancement both because they are only partially open to his influence and because their behaviour is to some extent open to historical variation, and is therefore unpredictable. On the other hand, he may find his bosses acceptable in so far as he has been able to come to terms with both himself and them, after long experience.

With this, the analysis of the worker's past and future, as he sees them, is concluded. Whether or not the reconstructions of covert behaviour that have been discussed are referred to the worker's 'personality' in the conventional sense (which could just as well include his judgement of his place in society), it is a reasonable finding that the two reconstructions are relevant only to his personal relations at work; though in two quite distinct ways, which depend on how personal these relations are. At all events, a relation with the worker's judgements of either his task or his pay could have been expected only on the remote and unlikely possibility that the person's character-type, as described by the psychoanalysts, uniquely determines his relations with objects, whether machines or currency notes. That this is not so does not preclude the possibility that some such minor relation would be found in a more varied group of workers, or simply in a larger group. The relations found, however, are more than marginal possibilities: they are clear and definite, and therefore require discussion at length. The last point that must be made in this discussion is, again, that the status of either of the generalizations proposed as a reconstruction of actual covert behaviour is more than doubtful. It may be that the worker's present account of his upbringing can be taken to stand for his habitual and recurrent expectations of other persons; but, if this is so, it can hardly refer to anything but persons in general. In the best case, that is, it might be taken to stand for his judgement of other persons in general and unexacting relations with him; it is very unlikely that it could be taken to signify his judgement of the particular persons with whom he has close relations. The worker's expectations for his future must equally be brought into the present, as a possible general judgement of his present situation. But, in this case, it is unlikely that the reconstruction set out above is a reasonable representation of the inarticulate behaviour that it purports to reproduce. The point has already been made, but it is worth repeating, that any worker presumably has some views on his possible future. The views set out above, however, are derived by remote reconstructions from certain limited information, and may therefore be much more coherent than those that might have been derived from a wider range of information.

It follows that what has been set out is only how the psychologist may laboriously reconstruct the way in which the worker sees his past and future. How the worker may see his present engagements and, in particular, what his hierarchy of personal values may be, remain to be considered in the next chapter.

CHAPTER IX

THE WORKER'S ENGAGEMENTS OUTSIDE WORK

THE suggestion made in the previous chapter is now taken up, that the worker's hierarchy of personal values may be related to his judgements of the work situation. That there is such a hierarchy is, of course, only a hypothesis. In simple terms, it would require only the arrangement in rank order of the person's main activities, according to their importance for him. What matters most to the person may be his work, his family (present family or family of origin), his friends (sub-classified in various groups) or himself: that is, the things he does by and for himself, for the sake of doing them. This simple statement, however, is not easily derived in practice. The direct approach used by Hammond (in Oeser and Hammond, 1954, pp. 246, 247) gives an inferred statement of the relative importance to the person of his work and his family. The inference needs checking against his behaviour, if only against that as he himself reports it; for a pair of hypothetical statements may reflect social stereotypes as much as they reflect the actual behaviour which follows from considered judgements of personal values. A weightier difficulty is that, in the present case, the hypothetical question of what the person would do if he had no need to work at all applies only to manual workers. It was found (cf. chapter VIII) that a majority of the men interviewed (factories A to E only) specified that they would go on with some kind of paying occupation. But it cannot be inferred that many of these, or even any at all, attached a high value to their present occupations. On the contrary: the minority (13 of 127) who specified merely that they would go on working (presumably in their present tasks) may have done so as much through sheer lack of imagination as because they find their work engrossing.

A less direct approach was tried (factory F, in 1951, only), in the use of an elaborate questionnaire which set out hypothetical conflicts between the person's main fields of activity, or engagements, and asked him to state which he would do, in the general case that he could do either of two things, but not both. This was found to be of little use, both because it required a working time out of proportion to its importance and because the answers had little internal consistency. It appeared that the situations put before the person placed an excessive strain on his imagination. This is not surprising, as the person's main engagements cannot be compared on a uniform basis, as if they all depended equally on his initiative. He must work for his living, and he can exercise his initiative only within the very narrow range of different employers who are prepared to engage persons of his particular trade or occupational level. He may marry and beget children, of his continuing initiative through a series of complex engagements. If he does, certain other responsibilities are inevitable in practice, even though they are not socially or legally compulsory beyond the bare minimum of his making some provision for his family. There is still a range of activities open to him which in one sense depends simply on his initiative. Thus he may put his major effort into family activities, into things he does with his friends or into things he does by and for himself. But his initiative is none the less limited, in two related respects. Nobody can do or take on more than time permits; and it follows that what anybody does take on, in the least prescribed or most open fields of activity, must depend to some extent on what he has taken on in other fields. For the family man, therefore, there is a lesser range of initiative than there is for the single man; and for any man the range of initiative is limited by the first requirement that he must work for his living.

As fields of activity, or major general engagements, cannot be compared on a reasonably imaginable basis, one must rely on the evidence of the person's reported behaviour. This is limited, in two respects. Firstly, there is no evidence of the person's behaviour at work. It may be that, even within the limited range taken, workers can be differentiated by the actual effort they put into their tasks; but it was not possible to investigate this. Secondly, the person's reported behaviour outside work is of variable reliability (as discussed below). What the married men do, in general, at home, can only be inferred from very general statements of habitual or recurrent practices. What the person does outside the home may be capable of exact specification as a crude fact, but it cannot be tied to the frequency or intensity of activity. In consequence, it may be possible to infer whether the person's family activities, or his activities with friends, or his self-directed activities, matter to him most. This is the utmost extent of general inference that can be made directly from the person's reports of his behaviour outside work. The importance to him of his work, beyond its being necessary to support his other activities, can only depend on more remote inference.

THE WORKER'S FAMILY

The primary facts are simply stated. Of the 127 men interviewed (in factories A to E only), 92 were married, and 77 of these had children, ranging in age from infants to independent adults.

Men who were separated or divorced from their wives and who had no effective contact with their children (if any) were classified as single; but three such men who were living with their mistress were classified as married. This simple classification shows which men have what sort of family responsibilities. How the married men carry out these responsibilities may be determined from their own statements of who decides about what, as summarized in Table 32. This derives from three distinct kinds of responsibility: who handles the family's cash; who decides what to do with the children (if any), in general and recurrently; and who decides on major but occasional projects, such as buying a refrigerator, car or house, or moving to another part of the country. The classification

TABLE 32

The division of responsibility in the family

Decisions are made by	Frequency
1. Husband and wife jointly, after discussion (which may include discussion with children, depending on the problem)	24
2. Either husband or wife alone, in largely separate spheres	24
3. The husband only or mainly	19
4. The wife only or mainly	25
Total	92

set out is simple and self-evident, but none the less requires comment. It derives from limited information; that is, from the person's brief statement of how he sees himself in action at home in certain specified respects. It is therefore doubtful whether this classification can be taken to be a description of behaviour at home. It is certainly not comparable to the classification of family types or structures set out by Herbst (in Oeser and Hammond, 1954, chs. X to XII). For one thing, the families considered cannot be compared. Herbst's account refers only to families in which at least one child was in his later school years. The present account includes families with no children and with independent children, or the whole range from newly married couples through to grandparents, including some old childless married persons. Even though it was found that there is no relation between the classification of responsibilities set out in Table 32 and the family situation (that is, age of children), it cannot be assumed that this classification can be compared with one which derives from a much narrower range of situations. A more important point is that the information gathered differs markedly in the two cases. Herbst's account depends on very detailed information given by the children themselves, who were naïve observers, although they cannot be regarded as being detached from the situation. The present account depends on information given by persons who were closely and directly involved in the family, as active partners; and the information gathered was not at all detailed, but included brief and summary statements only. It would therefore be very rash to take the summary of these summary statements as a description of actual behaviour at home; but it may reasonably represent the person's general view of himself in action at home, and therefore may give some indication of the importance to him of his home and family.

This last possibility must be elucidated by reference to the classification made in Table 32. The men placed in category 1 are those who clearly and definitely state that all three kinds of family activities (spending money currently, the general treatment of the children and special occasional projects) depend on joint decision after discussion, which may include the older children if the matter under discussion requires that. These men may be called co-operative, to the fullest extent. In category 2 there are the men who make their own decisions in some fields of activity (mostly current expenditure) but who make no decisions at all in the other fields, which are managed by the wife or mother of the family (mostly the general treatment of the children). In this category, however, there is a considerable overlap of the responsibilities assumed in the third field, that of the special and occasional events or changes in engagements. The men in this category may therefore be regarded as co-operative to a limited extent. The men in category 3 could be regarded simply as autocrats, but this would not be an adequate representation of the interview records. From the records, the only necessary implication for a man placed in this category is that both he and his wife find it convenient that he should make the major decisions (in particular, the decisions that involve spending money, whether currently or for special purposes). Both parties are then free to pursue their own purposes, within the limits set by the decisions made. A parallel case may be put for the men in category 4. From the records, it is clear that these are not necessarily men who are bossed by their wives. The only necessary implication is that the man is content to let his wife run the home and family, as an ongoing business, so long as he can largely ignore it and attend to his own affairs. The men in both categories 3 and 4 may therefore be placed together (as they are in the following analysis) on the ground that they are the least co-operative in family affairs, whether by their regulation of family affairs in order to leave them a wide area of free activity (category 3), or simply by their refusal of engagements in the family (category

4). The four categories of Table 32 therefore probably signify only degrees of co-operativeness, or involvement, in family affairs; and only three distinct degrees, as the last two categories are taken together.

It can be said, then, that some of the married men see their family activities as of central importance and put considerable effort into them (category 1 of Table 32); while at the other extreme some men see these activities as relatively unimportant and presumably put considerable effort into activities in other fields (categories 3 and 4 together). The most interesting possibility in the latter case is that the person's major effort goes into his job. If this possibility is to be tested by reference to the person's judgements of his situation at work, it must first be analysed carefully, in order to determine which of these judgements might be related to the person's behaviour outside work, or to the hierarchy of values which is inferred from his stated behaviour at home.

The most general possibility is that the person is passively unco-operative at home because he is not very interested in his family. He spends as much time at home as do more co-operative persons, but the centre of his interests is in the factory: the activities into which he is prepared to put an effort are his task, his relations with his bosses and his formal or informal relations with his mates. The direct test of this would be the determination, from systematic observation, of whether or not he does put more effort into these activities (and especially into the task) than do the persons who co-operate fully in family affairs. As there is no relevant information, the test cannot be made. A second and related, but very restricted possibility, is that the person is actively unco-operative at home because he spends as little time as he must at home and as much time as he can in the factory. It is conceivable that the worker might do the latter because he is absorbed in his task, or because he has identified himself with the values or expectations of his bosses. However, the possibility is sharply limited in practice. Beyond his 40-hour week, the worker can go on working only at overtime rates: and the possibility of overtime work (as noted in chapter IV) depends fully as much on the nature of the task and on the particular firm's requirements as it does on the worker's initiative, even though the possibility was to some extent open to most workers at the time of the investigation. The relation between the amount of overtime worked and the person's general reported behaviour at home was therefore not considered, because of this general limitation and also because of the unreliability of the information on the extent of overtime work. This restricted possibility may therefore be disregarded in practice. The third possibility is that the person is actively unco-operative at home because he puts considerable time and effort into activities that are derived from the factory but which are not directly connected with his task. For example, he may be an active member of the firm's social and sporting clubs (if there are any), or he may regularly join his mates in a beer-drinking session at the local hotel before he goes home. If he does either of these, it can be expected that his reported degree of co-operation in family affairs will be low.

These three possible modes of being unco-operative in family affairs can be summarized simply. The person may be unco-operative at home because he has a large or even an over-riding interest in his task; or because he is closely identified with his bosses and their expectations of him (which would probably be related to his absorption in the task); or because he is closely identified with his workmates. In the first case clearly, and in the second case possibly, he could be described as work-centred, in the sense that his work would appear to him to be his most important activity. In the third case, his central field of activity would be neither work nor home. All three cases could be described as particular varieties of self-centredness, as distinguished generally from the case that the married worker's centrally important engagement is with his family. In no case is there any workable possibility of determining the issue by reference to particular evidence of what the person does; e.g. how much effort he puts into his task, how closely he is related to his bosses, how much time he spends with his workmates outside the factory (beyond the bare fact of how many of them he sees outside the factory, as noted in chapter VI). But there should be some relation between the worker's judgement of his task, bosses and mates and his report of his degree of co-operativeness in home affairs. There is no reason to expect any relation with the worker's judgement of his pay, and the possible relation for the other three judgements would be subject to the unknown but perhaps not very close relation between these judgements and action in the work situation. None the less, some significant relation might be expected, for the persons who are hypothetically work-centred (as inferred from their lack of interest in home affairs) are presumably more likely to think well of their tasks than are the other persons; and so for the other judgements.

The finding, however, was that there is no relation at all between the person's reported degree of involvement in family affairs and any of his judgements of the work situation. If the person's central interest is in social activities which exclude his family, it is apparently not in activities with his workmates: a point which has already been made in chapter VI, but which may be turned into a positive form in the following discussion. Much the more important finding is that, if the person's central interests are outside his home, they are not in his work, whether in the task itself or in his relations with his bosses. To put it simply, the factory worker is not work-centred, whatever other central interests he may have: for him, work is merely an activity which is necessary to support his other and personally important activities, whatever they may be. However, it may be that the summary of reported behaviour at home gives no clear indication of the person's relative values. The negative finding is of importance if it can be relied upon; but its reliability can be determined only by a further analysis which may elucidate the nature of the worker's engagements outside his work.

THE WORKER'S UNPRESCRIBED ACTIVITIES

What the person does, apart from his work, can be specified in detail from the interview records. The range of information includes a variety of particular activities, which are recurrent or habitual in the same way as the person's reported general modes of family behaviour, but which are much more specific in that the activities reported can be referred to a number of distinct objectives or social processes. The whole range may be referred to as unprescribed activities because of the nature of its components, as shown in Table 33. Some of the activities listed there could be regarded conventionally as leisure or spare time occupations; but some could be so regarded only if the idea of leisure time or spare time is pressed beyond its reasonable limit to include the time which the person uses to get on with urgent and important activities for which he is not paid. A more intelligible distinction between the activities listed and the person's other activities depends on the distinction made above between what the person must do and what he may do, of his own initiative. He must work for his living: this is a prescribed activity. If he has a family, he must assume some responsibility for the persons who depend on him; so that some degree, if only a minimal degree, of co-operation is in this case prescribed. Beyond this degree, what he does at home depends simply on his initiative; and the same is true for what any person (married or single) does outside his home altogether. There is a wide range of activities, therefore, which are not prescribed at all, whether by law or by social custom. The whole range may therefore be called unprescribed activities; which term includes purely leisure activities equally with serious kinds of work which the person does for himself, without pay. As they are set out in Table 33, the range of reported activities of all 127 men (married and single) is considerable, both in frequency and in kind. If these activities are to be related to the married man's report of his general behaviour at home, or if they are to be used to give an alternative formulation of the person's relative values, the necessary first step is to reduce the whole set to a simple order. This was done by computing the associations among the 17 activities tabulated. Not many of these were significant, so that the grouping

TABLE 33

Unprescribed activities

The worker reports that he	Frequency
1. Goes to night school, for whatever purpose	9
2. Has a second paid job (night or weekend), or is building his own house	27
3. Does household duties: prepares meals, does shopping, maintenance or repairs, attends to garden, helps with children (e.g. baths, feeds, puts to bed)	72
4. Visits relations and friends, goes to parties and dances	89
5. Goes to the pictures, theatre, concerts	72
6. Goes swimming, shooting, rabbiting, fishing	23
7. Reads or hears on radio low-class material (thrillers, serials, etc.)	46
8. Goes to church, whatever denomination	17
9. Goes to trade union or political party meetings	17
10. Reads or hears on radio high-class material (plays, concerts, talks)	18
11. Goes to see any of: football, cricket, tennis, cycling, speedway racing, etc.	40
12. Plays any one of: football, cricket, tennis, golf, etc.	30
13. Attends "men only" socials or clubs, goes to lodge meetings, is committee member	22
14. Engages in formal social activities: club or band meetings (open: not male only), goes to horse-races	27
15. Makes things, in workshop: as hobby, not necessity	20
16. Bets, regular or occasional (off-course, cf. category 14)	36
17. Drinks, regular or occasional	73

of activities was carried out on a somewhat forced basis. The result, however, was clear. The final grouping, as set out in Table 33, includes 7 family activities (1 to 7), 3 intellectual activities (8 to 10) and 7 conventional activities of the Australian male (11 to 17). These groups are distinguished on empirical grounds, but the distinctions can be supported by reference to the interview records. Categories 1 and 2 include only family men who are working hard to improve their home circumstances. Category 3 obviously includes only family activities. Categories 4 and 5 refer mostly to men whose social activities, as specified, are undertaken with their families (wife or children or both). The activities in category 6, again, are usually a family affair; there are family sporting activities (as specified) and distinctively male sporting activities, as in categories 11 and 12. Category 7, lastly, includes what the person does at home when he is clear of all other engagements: he sits down, relaxes and passively absorbs entertainment. The first 7 categories of Table 33, then, include a variety of activities; but all have the common feature that they are likely to be undertaken for the sake of the family, or in the family context. In this broad sense, all can be regarded as family activities.

The relatively infrequent activities in categories 8 to 10 fall in a different 'dimension', in that they are likely to be undertaken equally by family men or single men. Their common feature is that the person looks beyond his immediate day-to-day situation and takes an active interest in the society or in the universe of which he is a member. These may therefore be called intellectual activities, as a sufficient distinction from the other groups of activities: it is in no way implied that the person's interest in these activities is purely or mainly intellectual.

The activities in categories 11 to 17 are those that would be regarded conventionally as purely or mainly male activities. It cannot be said that this generalization is supported by particular evidence, for the evidence about women workers (which is set out in chapter X) is for a group which is not directly comparable with the men with regard to unprescribed activities, as a very large number of the women interviewed were young and single. Even so, the differences found (as noted in chapter X) are sufficient to make the generalization proposed plausible. Organized sporting activities (categories 11 and 12) are likely to be undertaken by men, whereas informal sports (category 6) are open to the whole family indifferently. Formal and organized social activities, similarly (categories 13 and 14), may be contrasted with informal activities (categories 4 and 5). Making things (category 15) is clearly a male activity in its *locus*: the woman who knits or sews does so in the house, in the company of her family, whereas the man who makes furniture or models or machines does so alone in his detached workshop. Lastly, betting and drinking (categories 16 and 17) are clearly predominantly male activities. To reinforce this point, it must be noted that these categories refer for the most part to the distinctively Australian institutions of illegal but popular off-the-course betting and of heavy drinking sessions immediately after finishing work (in Victoria, the hotels close at 6 p.m.).

It may be taken, then, that the person's unprescribed activities are sufficiently specified in detail in the categories of Table 33 and in general in the classification of these categories into three groups. The next question that must be considered is whether these activities are related to the person's family situation. To answer this question, one may reasonably include all the men interviewed, not only the married men. The latter may be classified provisionally, as in Table 32, as co-operative for all purposes (category 1), co-operative for specific purposes only (category 2) and unco-operative, or not interested in family activities (categories 3 and 4). If the single men are added to this classification, it is reasonable to divide them according to their age, separating the older men who are not likely to marry from the younger who contemplate marriage or who are likely to undertake it. The division that was made was arbitrary, as there was no clear breaking point in the age distribution of the 35 single men. In consequence, they were divided simply for statistical convenience into older and younger halves. The person's unprescribed activities were reduced to order simply by counting them. Some of the activities noted could be classified only as engaged in or not engaged in: for categories 1, 2, 8, 15, 16 and 17 of Table 33, the only distinction that could be made was between any reported engagement, whatever its apparent extent, and none. For categories 5 and 6, the same rule was observed, on the empirical ground that very few men undertook more than one of the activities specified. In this case, therefore, the only reliable distinction was between any engagement and none. For the other nine categories, however, there was a clear and usable range in the number of activities undertaken: for example, the men placed in category 3 may undertake from 1 to 6 distinctly specified household duties, and the men placed in category 11 may at some time observe from 1 to 5 distinctly specified sports. The count of unprescribed activities, therefore, was not a simple count of the fields of activity reported: it included the total number of activities reported in some fields. Lastly, two separate counts were made. All activities were summed indiscriminately, to give an index of the person's total activity outside his work. The range for this index was from 17 to 1, with an average of 6·3. The other index was the sum of the person's family activities (categories 1 to 7 of Table 33), less the sum of his intellectual and male sporting activities (categories 8 to 17). This index may be taken to indicate the person's relative involvement in family affairs, on the ground that the three types of activity are distinct in their nature. The procedure adopted is supported by the observation that the scales for the three types of activity are independent: for the three associations in turn, $P = 0{\cdot}44$, $0{\cdot}78$ and

0·40 that the relations observed are merely sampling accidents. The range for this second index of unprescribed activities was from + 6 to − 8, with an average of + 0·7.

Both indices were related to the person's family situation, and the results of this analysis are shown in Table 34. From this table, it is clear that the person's total of unprescribed activities (index 1) is not intelligibly related to his family situation. The only intelligible division of the table for this index is between the married and single men (categories 1 to 3 and 4, 5, respectively), as it appears that the married men are more active. However, the difference between the two groups is insignificant (P = 0·11). The first conclusion is therefore that the person's total activity outside his work is not related to his family situation; and this may be reinforced by the observation that total activity (index 1) is not related to the balance of activities, or the difference between family activities and others (index 2): P = 0·62 that the relation between the indices is a sampling accident. In consequence, the total of the person's unprescribed activities may be disregarded. For the balance of his activities (index 2), the case is different. From Table 34, there is a clear and simple tendency for the men who are most co-operative at home to undertake family activities and, at the other extreme, for the young single men to undertake other kinds of activity. The simplest test of this tendency is the computation of the correlation ratio (η) which is 0·366 for all five categories of Table 34 and 0·323 for the first three categories only: and both these values are highly significant. The general conclusion is therefore that the married worker's statement of his habitual behaviour at home may be relied upon to indicate the importance which he attaches to his family engagements, because

TABLE 34

Family situation and unprescribed activities

The worker's situation is	Frequency	Average number of unprescribed activities	
		Index 1	Index 2
1. Married, and co-operative for all purposes	24	5·9	2·4
2. Married, co-operative for specific purposes only	24	7·0	1·3
3. Married, not interested in family activities	44	6·8	0·4
4. Single, old (age 27 or over)	17	5·8	− 0·2
5. Single, young (age 26 or under)	18	5·5	− 0·5
	127	6·3	0·7

it has a significant parallel in his detailed report of his unprescribed activities. The men who describe themselves as generally co-operative tend to be those who engage in family activities, while the men who describe themselves as not co-operative (presumably, not interested) tend to be those who engage in activities which exclude the family. The finding that the worker's degree of involvement in his family affairs is not related to his judgements of his situation at work may therefore be accepted at its face value: the worker's centrally important engagements may be with his family or with himself, but not with his job.

It would be rash, however, to propose this conclusion without further analysis. Among the married men, there is a wide range of ages, from 22 to 67; and a wide range of family responsibilities, from the case that the young married man has no children yet through to the case that the oldest workers are grandfathers. It is at least possible that the married worker's description of his general behaviour at home, or his statement of his unprescribed activities, depend simply on his situation or particular responsibilities or on his general experience or age. These possibilities must be considered before it can be concluded that the general description of behaviour at home can be used for the inference of relative personal values.

It has been noted already that there is no relation between the person's general description of his behaviour at home (or his hypothetical degree of co-operation in family affairs) and the age of his children. The person may be co-operative or unco-operative regardless of whether he has no children or (at the other extreme) he has independent children who are themselves married. The relevant second finding is that there is no relation between age and degree of co-operativeness: for the married men, co-operative behaviour does not depend at all on how old they are, in spite of the wide range of ages.

The person's unprescribed activities require separate consideration. That the sum of these (family activities, less intellectual and sporting activities) is significantly related to the person's statement of his general behaviour at home, is interesting. But the relation is only statistically significant. It is far from being close or definitive, so that the person's unprescribed activities equally may depend on his situation or his experience, or both. In this case too the findings are clear. The sum of family activities (index 2 in Table 34) is not related to the person's family situation (from no children through to independent children), or to his age. There is a possible link between the family situation and the relation between family activities and general behaviour at home. For all the 92 married men, the correlation ratio (η) for the scaled sum of activities with the three general categories of co-operativeness is 0·323 (as noted above). For the 30 men who have no children or independent children, the value of η is 0·486. For the 62 men who have dependent children (infants, or of school age), $\eta = 0{\cdot}275$: and the difference between the two figures is significant. It may be, then, that the person's differential total of family activities depends to some extent on the possibilities of his situation. He can perhaps undertake the family activities listed in Table 33 more easily if he has no children or independent children: if he has dependent children, he is to some extent tied to fundamental supporting activities which are not differentiable in the terms of this table. The scope of the person's initiative, that is, may depend significantly on the size of his immediate responsibilities. This is not a remarkable conclusion, and it only reinforces the previous conclusion that the person's report of his general behaviour at home may reflect the order of importance to him of his present engagements. What he actually does outside his work may depend, variously and complexly, on his circumstances; but, if the circumstances are disregarded, it seems likely that reported behaviour in particular (that is, the balance of unprescribed activities) is derived from and reflects back into family behaviour in general. The last finding that need be stated, then, is that there is no relation at all between the balance of unprescribed activities (index 2, for all 127 men) and the judgements made of the work situation. The balance of the person's unprescribed activities is as irrelevant to his work as is his general behaviour at home. Both statements of the person's behaviour outside work may be taken to reflect the relative values that he finds in his engagement with his family and in his various self-centred engagements (including activities with his friends, intellectual activities and purely personal activities).

CONCLUSION

The general conclusion is easily stated. Consistently negative findings are quite as important as positive findings, though they can be set out much more briefly. There is no relation between the worker's evaluation of the engagements that depend to some extent on his initiative and his judgements of his situation at work. His statements of his general behaviour at home and of his detailed behaviour outside work are coherent, intelligible, and intelligibly related to each other. It can be concluded, therefore, that there are some married workers whose most important engagement, as they see it, is with their family and others whose most important engagement is with themselves (in various distinct respects). This very broad generalization must be taken with reservations. For one thing, the person's involvement with his family or with himself only ranges by continuous degrees between the extremes. A more important point is that the generalization depends on a particular and limited range of information. In general, it should be possible to distinguish a number of centrally important engagements (such as intellectual, or tribal-totemic or purely autistic-constructive); but these, on the basis of the information available, must be regarded indiscriminately as the person's activities by and for himself.

None the less, the main generalization proposed can be supported indirectly by the consideration of the relation between the person's present engagements, other than work, and his report of his upbringing. It was observed, in chapter VIII, that there is a highly significant relation between this latter report (as detailed in Table 30) and the worker's judgement of his mates. The conclusion was that upbringing, as reported, may be taken to stand for a general present judgement of personal relations. The report may be a re-valuation of early experience, in the light of later experience; but, whatever its basis may be, it is apparently extended to include the most general and the least exacting personal relations, the relations of the person with his workmates. It may be suggested now that this is the only case which is likely to be capable of generalization from early experience, whether or not that experience is successively re-valued in the light of later and more specific experiences of personal relations. The relations considered in chapter VIII are dissimilar: there is no obvious continuity between the child's relation with his parents and the adult worker's relation with his peers (i.e. his workmates). But the person's present social relations outside work are even more different from his early family relations. The transfer from the situation of the child with his parents to that of the father with his wife and children requires a reversal of role rather than a mere transfer. The child's situation is relatively unprescribed or, at the least, unexacting. The situation of the husband or father is exacting because it is a close and continuous personal relation which he has entered of his own initiative. But, because he has done this, the relation is open to experiment: it is likely to depend to a considerable extent on the person's learning how to handle particular situations, and especially how to get on with particular individuals to whom he is closely committed, but whose ongoing relation to himself can be determined only by action that is directed with both love and intelligence. A simple relation between reported upbringing and present family engagements is therefore unlikely.

The case with the person's unprescribed activities is a little more complicated, as these activities are somewhat various. Some are not social activities, but purely the activities of the person by and for himself. Some are family activities, whether directly or indirectly (mostly the latter). Some, again, are social activities outside the family, whether informal (such as drinking with friends) or formal (such as playing in a football team or in a band). The two latter kinds of activity are both exacting; the family activities, obviously, in the same way as the person's general engagements with his family, and the social activities because they depend on the informal expectations of other individuals or on the formal expectations of institutional groups. In general, then, the person's unprescribed activities may depend considerably on the person's learning how to handle particular ongoing situations; but some of these situations (the person's individual activities) do not depend on other persons at all. A simple relation between reported upbring and present unprescribed activities is therefore even less likely than is the possible relation between upbringing and present family engagements.

The findings support both of these hypotheses. There is a just significant relation ($P = 0{\cdot}024$) between upbringing and present family engagements. The married men who grew up in normal and happy homes (categories 1 to 4 of Table 30, in chapter VIII) tend to be those who are most co-operative in their present family (category 1 of Table 32), while the other men tend to be less co-operative (categories 2 to 4 of Table 32). The child's experience is not directly of co-operation or lack of co-operation, but only of reliable love, or unreliable love, or reliable lack of love, in his dependent position. The transfer of this experience, or even of the eventual re-valuation of this experience, to the situation of adult responsibility in the family is clearly dependent on much intervening experience, and also on the particular persons in the present situation. The person's unprescribed activities (index 2, or the relative total of family activities) were found to have no relation at all to his upbringing. In this case, it would appear that the person's present behaviour (as he reports it) is too remote altogether from his earliest experiences, and is too dependent on his present circumstances and possibilities, to be capable of derivation from the first stage of his history, even as that is re-valued in his present report.

These findings to some extent support the idea put forward in chapter VIII, that the person's present report of his upbringing is a re-valuation of his past in terms of his experience of persons in general, and is therefore relevant only to the person's most undemanding relations: his relations with his workmates, in which he need not assume any more personal responsibility than that of the child in the family, though on the quite different ground that he is relatively independent of his mates. The point made in chapter VIII may be repeated, that there is a certain parallel in the early family situation, in so far as the child's relations with his siblings are less exacting than are his relations with his parents. The findings equally support the idea (if only by indirect inference) that the worker's statement of his general behaviour at home can be taken as a reasonable indication of the relative values he attaches to his various engagements; subject only to the reservation that the only clear distinction is between the persons whose centrally important engagement is with their family and the persons whose various central engagements are with themselves, in different respects which cannot be reliably distinguished.

The most important point, however, is obvious. There is nobody who sees his work as his most important engagement. The worker may be family-centred or self-centred, depending somewhat on his situation; but he is never work-centred. This, again, is an obvious conclusion. It is a convention of applied psychology that there are persons who regard their work as their most important engagement and who therefore put their heart, soul and guts into creative activity at work. But the possibility of doing this is open only to a very few professional or independent workers. Most workers of any grade have no such scope, and factory workers certainly have no scope at all for creative activity: they do what is prescribed exactly by the task itself, and can do neither less (unless they are prepared to be dismissed) nor more. For the factory worker, work can only be the means of earning a living: a task which is more or less acceptable, as argued in the preceding chapters, but which is only accessory to the main purposes of life. The engagements into which the factory worker puts his major effort, as judged by the reconstruction of his hierarchy of values, may vary considerably; but, as with nearly all persons in other occupational grades, his major effort goes into activities outside work.

CHAPTER X

THE WORKER AND HIS CONTEXT

THE analysis of the worker's judgements of his situation at work has now been taken almost to the possible limit, and conclusions have been stated about the relevance to these judgements of the situation itself, of the worker's history and individual engagements, and of his place in society. The order may be reversed: the worker's judgements of his work apparently depend on social structure and personality in the factory. But the dependence is in some respects by no means obvious because some, at least, of the conclusions stated are forced and speculative. This critical point must be considered in detail in due course (in the last chapter): meantime, the conclusions may be indirectly supported by the results of certain comparative analyses. The two distinct sets of evidence are considered in turn.

Firstly, the men who were interviewed may be compared with women who work in factories and with bosses of two grades, immediate supervisors and professional workers. The details of the numbers of persons interviewed in either group were set out in chapter II and further details are stated below. The relevance of this information may be limited in practice, but it is clear in principle. If the male worker's judgements of his situation as a worker do really depend on his actual situation (among other factors), then the judgements made by other persons (women workers or bosses) should differ intelligibly, as their situations differ. However limited the comparison may be for practical reasons, it is clearly worth making in principle. The second kind of comparative analysis, by factories, is of a quite different kind and is less obviously relevant to the general findings proposed so far. Any factory is a social organization which has individual and distinctive characteristics, if only because it is run by a small group of persons whose individuality (or singular characteristics) may be as important as are the tasks that they have in common with all other directors, managers and supervisors of factories. It is conventional in industrial psychology that the management of the particular factory is important; one could say, indeed, that industrial psychology is the practical study of the relation between management and the worker, with particular reference to the worker's productivity (or efficiency in general) or, at the least, with reference to his satisfactions in work. Industrial psychology, that is to say, is an applied art rather than a body of exact and general knowledge. It has been concerned, for the most part, with the possibility of increasing efficiency (or productivity) by the use of efficient methods of worker selection and training. A newer but related concern is the field of 'ergonomics' in general: the worker's efficiency may depend to a significant extent on the physical conditions of his task, which are therefore studied from this point of view. Lastly, the implication of social psychology applied to industry is that the worker's efficiency may depend on his relations with other persons at work, and especially on his relations with the managers of his factory.

Whatever the possibilities may be, the conclusion is clear. Factories do differ in the methods of their managements, as well as in the tasks that they require to be done; and the differences among the seven factories studied can be specified in reasonable detail. These differences cannot be related to differences in working efficiency, because the latter was outside the scope of an investigation in general terms and not merely in industrial psychology; but they can be related to the person's general judgements of his situations at work. In consequence, it should be possible to clarify, if not to define, the extent to which these judgements depend on the worker's situation in particular. That they do depend on his particular task, the system under which he is paid, and his actual bosses, has been shown. The question, then, is how far these elements of the worker's situation depend on the individuality of the factory: it may be that the worker's situation is simply the industrial situation at large, or the general situation of all workers, or it may be that his local situation, in a particular factory, is important. So far as this question can be determined, it is relevant to the question of the generality or reliability of the findings reported above, though less directly than is the comparison of workers in obviously different situations (men, women and bosses). Further, it may be possible to make some general statement of the practical importance of differences between factories. In one way, this would be directly related to the general aim of the project, the investigation of industrial tensions, or of disagreements between management and workers. As was explained in chapter I, it was thought appropriate to refer this aim to the general study of behaviour at work (that is, of judgements of the work situation), rather than to make another study of the apparent local determinants of efficiency. However, so far as local and practical conclusions can be stated, they are appropriately stated in this chapter.

MEN, WOMEN AND BOSSES

The comparison of men with others can be made only on a limited and doubtful basis. It was noted in chapter II that in two of seven factories only the men who were interviewed were representative of the whole body of men with respect to sections of the factory, age and length of service. In four other factories, representativeness was doubtful; and in factory G there was clearly no adequate representation. This last point may be related to the observations in chapter VI on the special derivation, and consequent special features, of the men who were interviewed in factory G (cf. also the second section of this chapter). However, it does not seem necessary to conclude that the men who were interviewed must be regarded as entirely unrepresentative. For the men, factory F must certainly be excluded in this section (it is noticed briefly in the next section), because the eight men who were interviewed here, in a predominantly female factory, were very highly privileged compared to the men who were interviewed anywhere else. It follows that conclusions in this section must be limited to 189 men (factories A to E, and G, as in chapter III). But this basis can be accepted, because the findings reported below do not differ if the basis taken is factories A to E only, or 127 men: so that the larger number may be preferred on statistical grounds. Again, it is true (cf. chapter II) that, because the men who were interviewed selected themselves, there is no possible check in any case on the possibility of gross bias by mentality or general outlook. On the other hand, some of these men (127, in factories A to E) seem to be closely similar to other and more certainly representative industrial samples, in the characteristics which are detailed in chapter VII. It may be concluded, then, that the men could be reasonably compared with an equally general group of women or bosses, even though the men (who were certainly the most representative of the three groups) can hardly be taken to be representative of industrial workers in general.

With the women interviewed, the possibilities are much more limited. The interview records for women, as noted in chapter II, are mostly not comparable with those for men, for reasons of detail which need not be repeated. There is therefore no possibility of analysing the women's judgements of the work situation in detail, as with the men. Even the analysis of the women's judgements in relation to the factory situation must be set aside as too speculative because of the very narrow range of tasks among the women who were interviewed. The most that can be reasonably proposed is the comparison of the judgements made by men and women. For this limited purpose, all 85 records of interviews with women can be used. More than a half of these (cf. Table 1, in chapter II) came from factory F, which is probably less representative of its industry than are the other factories; but, with regard to the judgements of the work situation, there were no differences among the four factories in which women were interviewed. It has been noted already (in chapter II) that both the men and the women who were interviewed range widely in age, length of service, nationality and occupational grade; so that both could possibly be regarded as general industrial groups, regardless of their particular location in certain not very representative factories. If this is so, the two groups would presumably differ, at least in those respects that one might expect to be determined by the worker's position in society as a man (conventionally, a breadwinner) or a woman (conventionally, a temporary or auxiliary worker).

In fact, the only differences noted are of this simple sociological order. The women interviewed were somewhat younger than the men (four and a half years, on the average), and were more often junior workers (aged less than 20), both differences being statistically significant. There were no differences by length of service, except that fewer women had very long service (15 years or more: which is beyond the range of Table 4, in chapter II). There were no differences at all by nationality; and the difference by occupational grade (cf. Table 6 in chapter II) was obvious, in that the women interviewed were more often process workers while the men, by comparison, tended to be either tradesmen or labourers. These few differences merely reflect the fact that women in industry are only temporary or auxiliary workers. It may be concluded, therefore, that the woman's situation at work can be reasonably compared with the man's.

For the bosses interviewed, the possibilities are even more limited. There is complete information on the general background of 29 bosses only (cf. Table 1, in chapter II). The information on the boss's situation at work is less complete than it is for the men, because it was even more difficult with supervisors of any grade to observe their behaviour at work than it was with their men. Further, the bosses' judgements of their situation at work cannot be systematically analysed; all bosses were interviewed primarily as consultants, so that specific judgements were not asked for. The consequence is that only the boss's judgement of his task is usable in a comparative analysis because it can be inferred (where it is not directly stated) from the records of the 29 bosses who were fully interviewed, and from 3 other records, making a total of 32. The most important point, however, is that the comparative analysis must be confined to this one point because the few bosses who were interviewed were obviously in no way representative of first line supervisors in general. The aim of the investigation was to include the immediate boss's account of his behaviour as a boss wherever possible, and then to compare this with the judgements made by his workers; and the results of this analysis have been reported in chapter VI. There was never any question of attempting a representative sample of junior bosses, because there was obviously no possibility of doing this. Far from being representative, therefore, the junior bosses whose interview records are usable even for the single comparison noted above constitute a very particular group which depends heavily on the circumstances of the investigation, as can be seen from Table 35.

TABLE 35

Characteristics of junior bosses interviewed

The boss's previous occupational level was	The boss's present level is	
	Professional (chemist)	Technical or supervisory
1. Professional throughout *or* supervisory throughout	13	4
2. Manual worker, *then* professional or supervisory	2	13
Total	15	17

In the first place, it is obvious that the division between professional bosses and supervisors (any or all of whom may be reported by the worker as being his immediate boss: cf. chapter VI) depends simply on the range of the investigation. There are few industries in which any worker could regard a professionally qualified supervisor as his immediate boss, so that the observed frequency of professional men (15 of 32) must be regarded as entirely unrepresentative. Secondly, there is an obvious link between the boss's present grade and his history, and it is statistically significant ($P = 0{\cdot}001$). The tendency is for professional bosses to have been professional workers previously (only 2 of 13 are exceptions); and for technical bosses (e.g. persons with some qualifications in engineering, but below the professional level) to have been manual workers firstly (there are only 4 exceptions among 17). It is clear that these 32 bosses cannot be treated as a homogeneous group; which conclusion only reflects the circumstances of the investigation. The consequence is that, although complete information (within the limits noted above) is available for 29 of the 32 men who are entered in Table 35, it cannot be used in analysis for lack of a suitable basis of comparison. Besides the findings noted in chapter VI, the only findings that can be reasonably inferred with regard to the junior bosses are those for their judgements of their tasks: for it can be assumed that any boss has a task which is superior to (more satisfying than) that of any of his workers. The point is worth repeating, however, that the group of bosses ranges widely in age and length of service (as noted in chapter II), although it is not a homogeneous group. At the least, it can be taken as a group of low-level bosses who are settled in their work.

The detailed comparisons, then, are limited to the four judgements of the work situation; and of these the first only (the task) can be referred to all three groups of workers. This first comparison is set out in detail in Table 36, in which the categories of Table 7 (in chapter III) are repeated. For the dichotomy between satisfaction and absence of satisfaction with the task (between categories 3 and 4 of Table 36), there is no significant difference between the men and the bosses. The women, however, are significantly more satisfied with their tasks in general than are either the men or the bosses. Further, among the satisfied workers (categories 1 to 3 only), there are clear differences. The men have detailed grounds for their judgement of the task (categories 1 and 2), while the women tend to approve the task regardless of its particular conditions (category 3). The bosses, by contrast with either the men or the women, tend to find their task varied and interesting (category 1 only). Again, there is a difference among the workers who do not report positive satisfaction with their task. For the men and women, the difference is insignificant. For the bosses, the obvious difference is that the chief attraction of the job may be not in the task itself but elsewhere (category 4, compared with category 5). In this respect, the bosses differ significantly from both the men and the women. The comparative conclusions are simply stated, on the basis of these findings. For the bosses, the task itself may be acceptable because it is varied and interesting; or the whole work situation may be accepted, regardless of the particular task, because the pay is good and the higher bosses are good. For the women, the task is accepted so long as it is within their scope: whether or not it has other hypothetically desirable features is not relevant. For the men, the case is intermediate between these two extremes: the task is judged as being satisfactory if (in comparative terms) it has compensatory or alleviating features. These conclusions may be regarded with scepticism, in view of the reservations noted above. For a representative group of foremen, the differences noted might well be less marked, or insignificant; but the point can still be made that the middle-class boss, even if he has only junior status, is more likely to find his task interesting (category 1 of Table 36) or irrelevant because it is overshadowed by other features of his work situation (category 4) than is the worker whom he bosses. For the women, the case is different. A half of the women who were interviewed were in factory F, and most of these were of middle class (or white collar class) origin, as was appropriate to the particular circumstances of this factory. But this origin is extremely unlikely for working women in

TABLE 36

Comments on their tasks by men, women, and bosses

The task is	Percentage incidence for		
	Men (N = 189)	Women (N = 85)	Bosses (N = 32)
1. Varied, interesting; good conditions, amenities, may also be mentioned	19	16	57
2. Easy, has good conditions, amenities: good pay or boss may also be mentioned	25	19	3
3. Excellent, very good, good, O.K., no further detail	26	52	6
4. Not mentioned: the *job* is good because of good pay, bosses, etc.	7	1	28
5. Dull, dirty, hard, tedious, boring: bosses, pay, etc., may be mentioned, as good or bad	23	12	6

general. It is therefore likely that women factory workers in general are more content with their usually inferior tasks (that is, dull and unskilled tasks: category 3 of Table 36) than are the men. In this respect, the differences noted are probably an underestimate of those that would be found generally. The general conclusions may therefore be proposed. For the junior boss, the task may be to some extent creative, depending on his day-to-day initiative, open to some extent to his initiative in development, and with possibilities for a career; subject only to the qualification that this is more likely to be true for professional workers than it is to be true for first-line supervisors without professional qualifications. For the man who works in a factory, the task may be complex and, in minor respects only, creative; and it may have possibilities for a career, though rarely and doubtfully. For the women who works in a factory, the task is just a task: it is in accordance with the woman's place in Australian society that her industrial task in general should be relatively uncreative (that is, purely repetitious) and should offer no prospect of a career beyond that of continuing in the same task.

The judgements of pay made by men and women are summarized in Table 37 (which derives from Table 8 in chapter III). In this case, the differences are obvious. For the main division between satisfaction and dissatisfaction (between categories 2 and 3), the women are significantly more satisfied with their pay than are the men. There is no significant difference with regard to the two kinds of dissatisfaction (categories 3 and 4); but the women tend significantly to be highly satisfied with their pay (category 1), while the men (by contrast) tend to be satisfied only with certain reservations or qualifications (category 2). These findings are interesting because women are generally paid less than men.

TABLE 37

Comments on their pay by men and women

The pay is	Percentage incidence for	
	Men (N = 189)	Women (N = 85)
1. Terrific, excellent, too much, very good, good	34	61
2. Good, sufficient, considering: job is inferior *or* can make extra (overtime, etc.)	32	22
3. Not enough to live on: may also be called, not bad for job	19	8
4. Not enough for job, which is skilled, unpleasant, better paid elsewhere, etc.	15	9

A direct comparison is difficult, as there are not many jobs that are open equally to men and women; but, for these jobs, the women's rate of pay is usually three-quarters of the men's rate, or in a few cases, nine-tenths. Within the limits of the investigation, however, the indirect comparison is easy. Many of the women who were interviewed in factories C, E and G and a few of those who were interviewed in factory F were doing tasks that were no less arduous and no less skilled than those of the men whom they could see working close at hand, or, in some cases, whom they relied upon for the process next before or after their own operation in the production cycle. It must be presumed that these women (who surely knew that the men in question were paid substantially higher rates) could have made a realistic and informed comparison between the tasks if they had been inclined to do so. It can be concluded, therefore, that the woman worker accepts her inferior pay not only as an unalterable fact (as it is) but also as a natural and proper fact, which derives simply from her inferior social position in general. Of course the findings noted might derive from the fact that women generally have lesser financial responsibilities than men. But this is easily tested by reference to their reported responsibilities. Of the 85 women who were interviewed, 49 were single and were working only to support themselves (young, or 29 or less, 34; old, or 30 and over, 15); another 29 were married women who were working to increase the family income (young, 9; old, 20); and the last 7 were women who were working to support their children because their husbands could not or would not do so (chronic invalids, chronic alcoholics, and deserters: all the women concerned were aged 30 or over). Thus there was quite a variety of responsibilities, and the least expectation would be that the women with the heaviest degree, the breadwinners of necessity, would be less content with their pay than the others. In fact, there were no differences among the five categories (as stated

TABLE 38

Comments on their bosses by men and women

The immediate boss is	The higher bosses are	Percentage incidence for	
		Men (N = 189)	Women (N = 85)
1. Extra good, lets you alone, helpful, fair,	Nice chaps, lovely, gentlemen	34	41
2. nice chap	Unknown	13	27
3.	Some good, others not: too remote, inefficient, crooks, twisters	25	16
4. Pleasant but incompetent: unpleasant,	Nice chaps, lovely, gentlemen	7	1
5. crook, queer, moody,	Unknown	8	2
6. rotten, a pig	Some good, others not: too remote, inefficient, crooks, twisters	13	13

above, with two sub-divisions by age) with regard to the reported judgement of pay. It appears that the direst necessity does not affect the woman worker's tendency to accept her inferior pay for what it is; and the general conclusion therefore stands that inferior pay is accepted in so far as a generally inferior social status is accepted.

The comparison of the judgements made of bosses of all grades is shown in Table 38 (which derives from Table 9 in chapter III). For satisfaction in general (category 1 only), there is no significant difference between men and women. There is a slight but significant tendency ($P < 0{\cdot}05$) for the women to think better of their immediate bosses (categories 1 to 3) than do the men. For the higher bosses alone, there are no differences. For the division between categories 2 and 3 (which may stand for the distinction between effective satisfaction and all varieties of positive dissatisfaction), the difference is significant ($P = 0{\cdot}003$). The women tend, on this definition, to be more satisfied than the men. The conclusion is therefore much the same as before. The women workers tend, on the whole, to accept their inferior position in industry and therefore tend to a more favourable view of their immediate bosses. It must be noted, however, that this conclusion might be due only to the behaviour of the immediate bosses, on which there is no information that is suitable for analysis.

The last comparison, of judgements of the person's workmates, is detailed in Table 39 (which derives from Table 10 in chapter

TABLE 39

Comments on their mates by men and women

The worker's relations are	Percentage incidence for	
	Men (N = 189)	Women (N = 85)
1. Excellent, very good, matey, good, all right, no snags	57	70
2. No strife: not bad, but prefers to have little to do with mates	12	14
3. Good, but poor with cliques or individuals (foreigners, old hands, etc.): or poor generally	31	16

III). In this case, there is no significant difference for the main division (between categories 1 and 2): the women are as likely to be on positively good terms with their workmates as are the men. For the other possible division (between categories 2 and 3), there is a significant difference ($P = 0{\cdot}018$): the women are less inclined to make definitely negative judgements of their workmates. The conclusion, therefore, is that women workers are somewhat more easily content with their workmates than are men. Although the difference is not at all marked, it can be related to, or referred to, general differences in the work situation, as with the other judgements of this situation. The working woman is less committed than the working man. Only a minority of the women work to support themselves until the age of retirement (that is, the older single women), and only a smaller minority (those who are breadwinners of necessity) work to support their families; but most of the men work to support their wives and families, and not for themselves alone. For the women, therefore, work may be a temporary and unimportant engagement; so that other workers may be accepted without argument or discrimination, at their face value. For the men, work is a necessary and fundamental engagement, so that other workers may be judged more critically. Even if the worker can largely ignore his mates (in some circumstances), he is not likely to be indifferent to them because he has to go on working with them, simply as a condition of his employment. Undemanding as the relation may be, compared to other relations in the whole field of work, it is likely to be more demanding for men than it is for women.

The analysis can be taken no further. For the reasons stated above, it is not possible to trace the origin or derivation of the judgements made by either the women workers or the bosses. The general distinctions, however, are clear. The bosses are relatively highly satisfied with their tasks because these are, on the whole, creative. The women workers are, on the whole, satisfied with their task, pay, bosses and mates because they have less at stake than the men and are therefore inclined to accept their inferior position in industry. The men make their judgements of the work situation within the limits which they know and accept as a matter of course. They must work but cannot work creatively; in consequence, their judgements can only be referred to particular situations that are known from previous experience (tasks, pay, bosses or mates) or that are possible in that the person has entered them of his own decision (occupational levels or classes that were, in principle, open to the worker in his earlier years). This very broad conclusion is not directly relevant to the preceding analysis of the male worker's judgements of his situation. However, the conclusion sets the limits within which the worker's judgements of his situation should be analysed; and the analysis reported (in chapters IV to IX) may be taken to be credible or reliable because it was made within these narrow limits.

MANAGEMENT AND THE WORKER

The way is now clear to return to the male worker's judgements of his work situation and to relate these to the factory as an individual and particular social organization. This analysis may be only indirectly relevant to the general argument; but it is relevant in practice because the tendency of most recent work in industrial psychology is to show or, at the least, to assert, that the worker's behaviour depends critically on the particular mode of management of his factory. That the investigation was directed to the study of disagreements (tensions) between managers and workers is immaterial because (as noted above) it was assumed that a general analysis would be more relevant than an analysis that was directed to particular differences between factories. None the less, these differences may be important. The possibility obviously depends on the scope of the investigation, and it must be made clear that the findings reported below are severely limited. They refer to seven factories only (factory F is included here); and to seven factories that were certainly not typical of Australian industry in general, simply because they were the factories in which the investigation was carried out. Beyond these seven factories there is the vast range of factories that were not investigated because the management was not interested or (at the extreme) were not even approached because it seemed unlikely that the management would be interested. It must be assumed, therefore (as noted in chapter II), that comparisons between factories can be made only in the very narrow range of those whose managements were well disposed towards industrial psychology; and these seven must therefore be regarded as particular cases that were among the more intelligently run or more progressive minority of factories in general. Limited as the range is, it is the range to which the worker's four judgements of his situation at work must be referred.

The first of these, the judgement of the task, is little related to the worker's situation in a particular factory. Satisfaction and dissatisfaction in general (categories 1 to 3 and 4, 5, respectively of Table 36) do not differ significantly among the seven factories; nor are there any differences among the categories of satisfaction (categories 1 to 3 only). It was found, however, that the men who were interviewed in factory G tended significantly to take a positively critical view of their tasks (category 5 of Table 36), while the men who were interviewed in the other six factories tended only to disregard the task in favour of other attractive elements in the work situation (category 4). It may be concluded that, from the worker's point of view, the task is likely to be accepted as it is. Its desirable or undesirable features depend much more on its general nature than they do on the initiative of the factory management.

The worker's judgement of his pay depends somewhat more on the particular factory. For the main division, which is made between categories 2 and 3 of Table 37, there are no significant differences among the seven factories. Among the satisfied workers, those in factories A and D stand out as being satisfied unconditionally (category 1), while those in the other factories tend to be satisfied only with reservations (category 2). This finding may be related to the observation that factories A and D offered more than the others in the way of extra payments. In A, about a third of the men were shift workers and several substantial marginal payments (as detailed in chapter V) were open to workers in any group; while in factory D a substantial production bonus was paid to all workers at the end of the financial year. Among the dissatisfied workers also there is a difference: the workers in factory G tended significantly to regard their pay as being not enough for the job (category 4 of Table 37). This finding may be related to the observation that factory G offered less than any other in extra payment of any kind, including overtime pay. However, the main finding is evidently as before (i.e. as in chapter V): it is the general possibility of relatively effortless margins above the prescribed rate that matters, rather than the details of the pay system of the particular factory.

The worker's judgement of his mates may be taken next and out of order, for two reasons. It is the last of the three judgements that requires consideration, as the worker's judgement of his bosses, in relation to the particular factory, has been reviewed in chapter VI. Besides this, it is important to repeat the point that was made in chapter VI, that the men who were interviewed in factory G were singularly unrepresentative. For these men, the finding is that they tended significantly ($P = 0{\cdot}003$) to get on better with their workmates than did the men in factories A to F. In one way it is preposterous to explain away this finding by using it only to support the point made both in chapter II and in chapter VI that in factory G in particular the men who were interviewed were quite unrepresentative. In another way, it is quite reasonable, on three grounds. In the first place, these men were grossly unrepresentative of the working sections of their factory, as noted in chapter II. Secondly, this unrepresentativeness may be related to the circumstances of the investigation, as noted also in chapter II, as well as in chapter VI. It is clear that many of the men who were interviewed were the foremen's nominees, rather than volunteers; and, in so far as these men came from certain sections of the factory only, it is likely that many of them were men who were on singularly good terms with each other. The only men who were drawn in directly by the investigators were a few of the maintenance tradesmen (total 20, of 62 who were interviewed in factory G), and the approach to these few (in casual off-duty conversation) had two distinct and obvious consequences: they persuaded some of their mates to come forward for interview, and their foremen then persuaded some others to go forward and present the other side of the picture. There is no doubt that the investigation in factory G was carried out as well as it could have been in the circumstances, nor that its execution was fully supported by the management. It is clear, however, that the men who were interviewed in this factory were the furthest removed of all from being representative. This does not depend only on the author's assertion: besides that it is shown by the particular points noted in chapter II, it might be expected, on general grounds, for any psychological investigation which is carried out solely for the purposes of the management. If for this last reason only, factory G must be distinguished from all the other factories, with the final note that the workers who were interviewed in factory G were probably disposed to take a better view of their situation in general than the view that would have been taken by their mates generally.

The third, and perhaps the most important, point is that there is no reason to expect the worker's judgement of his mates to differ from one factory to another. This is the one judgement which, if it does depend on the local situation (as could not be shown in chapter VI), can hardly depend on the management's initiative or on differences between factories and their managements. At the least, if the management selects workers with an eye to productivity, the workers like or dislike each other with an eye to human qualities that are not necessarily related at all to productivity. It is therefore likely that factory differences in the worker's judgement of his mates are merely sampling differences. The fact that the only difference found is between factory G and all others therefore only reinforces the conclusion suggested above. The men who were interviewed in factory G were a special set of close in-groups, mostly of long service, who were chosen by their foremen (mostly) or by their mates (partly). The fact that these men were on better terms with their mates than were the men who were interviewed in factories A to F can therefore be disregarded as being merely circumstantial and incidental.

This somewhat emphatic, if highly derivative, conclusion would tend to strengthen the conclusions reported above, that the workers in factory G tend to positive criticism of their tasks, and of their pay: if the workers who were interviewed were the more satisfied (on the whole, being on the whole the foremen's nominees), the views of the whole factory population would presumably have been even more critical. The same point presumably holds for the worker's judgement of his bosses. As this was examined in detail in chapter VI, the findings need not be repeated. It is sufficient to recall the main findings only. The judgement stated of certain intermediate bosses (general foremen) was significantly more favourable as stated by the men who worked under these bosses than as stated by the independent tradesmen who merely observed their behaviour. But all workers in factory G tended to be more critical of the top management than were workers in the other factories. If it is taken that the workers who were interviewed in factory G were largely the foremen's (and especially the general foreman's) nominees, the conclusions are obvious. As was suggested in chapter VI, a boss who is exceptionally bossy (within the limits of the investigation) may none the less be approved and accepted by a selected group of his older and more reliable workers. But it does not follow that even these workers will approve and accept the top management: in so far as this is not a group of known persons, but merely an anonymous collectivity, it may be disapproved more than in factories in which some or all of its members are known as individuals.

The general conclusion is clear. There is no evidence of the worker's behaviour at work; but his judgements of his situation clearly depend much more on the industrial situation in general than they do on particular local differences between factories. This may be only because the scope of the investigation was limited, or it may be because the scope of any factory management (which is concerned primarily with production) is limited. The possibilities are discussed below. The relevance of the general conclusion is equally clear. The findings reported in the preceding chapters show that the worker's judgement of his situation at work depend on the factory, on social structure (or the complex judgement of this), and on personality, in that order of importance. The obvious question that must be considered is whether the first term in this series of three is to be taken as the industrial situation in general, or as the local factory situation in particular.

A short answer to this question can be derived from the findings reported above. The seven factories can be ranked in order of effectiveness, on a scale which sums the worker's four judgements of his situation by allocating 2 points or none for major distinctions (between satisfaction and non-satisfaction, as defined in chapter III) and 1 point or none for minor distinctions (among kinds of satisfaction, or between neutral non-satisfaction and positive dissatisfaction). If this is done, it is found that the scale for the seven factories ranges from 6 to 2: but factories A to F are all at positions 5 or 6 on the scale and factory G alone is at the position 2. The details (which can be re-traced from the above account) are therefore not worth reporting. The main findings, however, are interesting. It is necessary to consider in turn why factory G should have a distinctive position and why factories A to F cannot be distinguished, in spite of their apparent individuality as social organizations.

For factory G, the most obvious possibility is that its distinctive position is merely a reflection of the special characteristics of the men who were interviewed; but this can be disregarded, in view of the findings. On the one hand, the men in factory G were less content with their tasks and their pay, in minor respects, and with their bosses as a whole. On the other hand, the evidence from the introduction of the investigation in this factory, from the finding that these men were on better terms with their mates than were men elsewhere, and from common observation during the investigation, all tends to the one conclusion: that a large majority of the men were the foremen's nominees. It seems unlikely that the foremen nominated their less contented workers; so it must be assumed that the differences noted would have been greater for a more representative group. In fact, the differences can be related quite simply to the observed features of the factory. The physical conditions of work and amenities were, on the whole, inferior in that there were more dirty, dusty or noisy work-locations than in other factories and the provision of lavatories, washrooms, canteens, lockers and so on was inadequate. The possibility of any worker's getting substantial payments above the legally prescribed rate was comparatively restricted. Lastly, the intermediate bosses (the general foremen) were on the whole exceptionally bossy and it was noticed that members of top management were comparatively rarely seen in the factory. It must be stressed that these observed differences were differences of degree only within the narrow range of the seven factories that were studied: the investigators' knowledge of other factories, together with the comments made by some workers on their previous jobs, make it clear that factory G is none the less properly placed among the more progressive factories. However, the conclusion is clear. The worker refers his judgements of his situation mainly or even solely to his present possibilities; if he finds that, with regard to the features noted by the detached observer, factory G is in some degree inferior to other factories to which he could conceivably or quite easily move, he is likely to be dissatisfied with his task, his pay, his bosses, or any combination of these. He may also draw the conclusion that the aspects of his situation to which he objects are due to the deliberate policy of the top management. This important point will be considered in due course. At this stage, it is sufficient to make the obvious point that the men in factory G tended to think poorly of their situation at work simply because this situation was inferior, in certain respects that could be specified exactly by the investigators.

The case with the other factories is not so clear because, from one point of view, all of these were markedly individual in their methods of management and general labour policies. Factory A is a recently established branch of a large overseas firm. Its management is fully up to date with American and British ideas of management and is careful to put these ideas into practice. The most obvious distinctive feature of this factory is that it has a Works Council (as it was called at the time of the investigation), or a joint consultative body, which is very rare in Australian industry; and many less obvious features could be listed to support the general distinction. Factory B, which is a branch of A, though older-established, can be included under this distinction, with a few minor qualifications due to its distance from the central management. Factory C is markedly different: it is, by origin, a purely Australian concern and, up to the time of the investigation, was a family firm; very well set up physically, but very easy-going and unsophisticated in its methods of management. Factory D is also a purely Australian concern and, by origin, a family firm. At the time of the investigation it was expanding rapidly and had considerably outgrown its site. The management had two distinctive ideas: that any worker who showed himself capable should be assisted by the firm to work his way up to an executive position (examples of this were noted by the investigators), and that the firm's annual profits should be divided equally between the shareholders and the workers. These two ideas may be summed neatly under the firm's practice of insisting that all members, from the managing director downwards, were equally workers. Factory E is a small purely Australian family firm which also, at the time of the investigation, had considerably outgrown its site, and which was distinguished by the friendly, not to say almost matey, relations between the owner-managers and the workers. Factory F is again an Australian family firm (though with overseas connections) and is exceptional in that it does not produce for the open market but only for the retail store which owns it. At the time of the investigation, the managers of all grades were at considerable pains to be on good terms personally with the workers, and the few men in the factory were paid well above the current rates, for what was a relatively unexacting task.

This brief and bald account could be greatly extended, to take in both the comparison of the factories' economic situations and the fascinating range of detail of policies and practices that was gathered by the investigators. The extension would be pointless, however. What has to be explained is the fact that the workers in the six factories A to F did not differ in their judgements of the work situation, so that there is no need to elaborate the point that these factories were in some respects distinct and highly individual, even more so than is shown by the brief statement made. The fact of distinct individuality is clear, but its apparent irrelevance to the worker's judgements is equally clear. The solution to this puzzle is simple. The observed individuality of the factory (including factory G, for most purposes) depends on the personal style of the group of top-level managers or directors. But this style is superimposed on the common ground of necessary industrial relations; and it is only the differences in the common ground that to some extent determine the worker's judgements of his situation. This may be demonstrated in concrete terms.

The worker's judgement of his task, firstly, depends on the actual conditions of the task, which (as was argued in chapter IV) he apparently assesses in much the same way as does the observer. Now these conditions are largely given in the nature of the task. Independence and completeness, certainly, have no conceivable relation to the style of the individual management; they depend simply on what is to be done. The cleanness of the task is to some extent, but only to a limited extent, open to the management's initiative. As the managements of all seven factories (including G) made considerable efforts, within the possible limits of their circumstances, to make the tasks of their workers as clean as possible, even this point must be disregarded as a ground of differentiation. Much the same holds for the worker's judgement of his pay. What matters to the worker is the possibility of relatively unearned extras, of whatever kind; that there were six distinct kinds (in factories A to F) of pay systems or policies was immaterial from the worker's point of view, so long as any of these gave the required result. In this case, there is clearly some scope for the management's initiative; but this is none the less limited to the first decision that substantial payments will be made. The worker's judgement of his bosses depends most clearly and simply on the management's initiative: it is simply a matter of policy whether the factory has officious bosses or co-operative bosses. This is true, but it is also beyond the range of the evidence available: apart from the variations in style that were noted above, there were no factories (including G) in which the top management were in favour of bossy bosses as a matter of policy.

A limited conclusion may be stated. The factory worker, whose situation is necessary to his living, does not see his situation in the same way as does the detached observer. The worker is not concerned with, or is probably quite unaware of, the finer points of differences in individual style that are apparent to the observer. If the task, pay and bosses are reasonably better than in other places which the worker knows or in other factories generally and hypothetically, that is sufficient: just how and why they are better is immaterial. Conversely, if the task, pay and bosses are worse than in other factories (on either comparison), that is the management's fault. This point is worth elaborating. The differences that were reported above, between factory G and factories A to F, are statistically significant but are of little practical consequence. None the less, it does matter that the task, pay and bosses were inferior (in specifiable respects, as detailed above) in factory G, from the point of view both of the workers and the observers. The points of view diverge, however, in explanation. For the observers, it was clear that the management of factory G had its individual style, certainly, but had the same general aims as the managements of other factories. The differences in the factory situation were clearly due to a number of historical and local factors (which need not be detailed, and which could hardly be detailed within the general requirements of anonymity), the total effect of which was that the top management of factory G had very much less scope for effective initiative than did their counterparts in any other of the six factories. The conclusion therefore stands: the worker judges his situation as he sees it now, as a permanent subordinate, and whatever more general judgements might be made (by management or by the detached observer) are simply irrelevant.

Even though this is a limited and provisional conclusion, it is drastic. The obvious point that must be considered next, before further conclusions are drawn, is that the limited conclusion derives simply from the restricted range of the investigation. There is no doubt that an investigation in a wider range of factories would have led to different conclusions; but the likely differences are easily specified if it is assumed that a wider range would have included not only more factories in general but a number of unpleasant, inferior, objectionable or unpopular factories in particular.

For the worker's judgements of his situation (as set out in chapter III), the likely differences are obvious. In a wider range of investigation, one would expect that fewer workers would be satisfied with their bosses, and also that fewer would be satisfied with their tasks and their pay, in so far as these are open to the management's initiative. The order of importance of the four elements in the work situation presumably would stand as in the present investigation, though other incidental and local elements might well assume a distinctive importance. For the local elements in the worker's judgements of his task, pay and bosses, the only likely case is that these would be more important or more effective in a more widely based survey, merely because a wider range of evidence would transform and extend the classifications that were made (in chapters IV to VI) of local conditions. The general elements in the worker's judgements (that is, his family situation, his perceived place in society and his personality) would not be likely to vary in nature or in detail in a wider range of evidence, though their relative importance might vary, either way. On the whole, it can be taken that the conclusions reported in chapters

III to IX would probably stand, in general terms, for a wider range of investigation, except that the importance of the local elements (chapters IV to VI) might be greater.

It seems likely, then, that what would matter to the industrial worker in general is his whole industrial situation, or at the least the general nature of his task, rather than the purely local features of his present situation in a particular factory. This implies only that large and general considerations come before small and particular; and the whole point of the present argument is that it may support the critical distinctions between large and small or general and particular. The limited and provisional conclusion can therefore be enlarged to a general and speculative form: the 'philosophy of management' is an empty gesture on the top of an intractable system of necessary relations. This conclusion properly emphasizes the point that the range of the factory management's initiative is limited; but it goes too far, by implying that there is no range at all. It is clear that the possible range depends on the elements in the worker's situation. To some extent, it is open to any factory management to improve the nature or conditions of the tasks that are required; but to a limited extent only, because some tasks are necessarily dirty (in relative terms) or dependent or incomplete, in their nature. If the worker's judgement of his task is taken to be the most important ground of his actions in general (and therefore, if only hypothetically, of his daily behaviour in the factory), it is important to note that the local grounds of this judgement are the least open to the management's initiative. The factory management may in part determine the worker's judgement of his pay (and so, hypothetically, determine in part his daily behaviour in the factory) by its decision on the range of extra payments that are available; but this decision is obviously limited, both by market considerations and by the management's responsibility to its shareholders. Lastly, it may be relatively easy for the management to initiate acceptable methods of supervision: but these will affect only a minor judgement among the four that are made by the worker.

Qualified as it is, the conclusion that the management's part in promoting worker satisfaction is necessarily limited by the nature of the industrial situation is still out of line with some conclusions in the field of applied industrial psychology. In one respect, the conclusion is obvious. The management's business is to get out production; and this is ultimately the worker's business too. If the particular interests of the two parties collide (and it is grossly obvious that they do so frequently, if not at all times), then the issue can be determined only by compromise. This simple point was made a generation ago by Roethlisberger and Dickson (1939, chs. XVIII to XXIII), who observed that output in some circumstances depended critically on the worker's assessment of its relevance to his purposes. The same point was made by Smith (1939), with reference to more pressing circumstances, and by Warner and Low (1947), with reference to a particular case of circumstantial stress. One may add to this the complementary views put from the worker's side. Barkin (1950), a trade union leader, observes that personnel management is irrelevant so far as it is concerned in principle only with getting out production. This concurs exactly with the point put by one of the tradesmen interviewed, a deliberate and dignified old man who had a variety of experience (over the span of two generations) in England, Scotland and Australia. His final comment was this: 'You hear so much talk of relations between the bosses and the workers. There'll never be peace between the two, they're both out for the most they can get, that's my opinion.'

In other respects, the general conclusion is by no means obvious because it contradicts the underlying assumption of industrial psychology in its most recent phase, that personal relations in the factory are the most important determinant of the worker's behaviour. The assumption, however, is disputable. In certain limiting cases, it is obviously true: if the management is fatuously incompetent, it surely invites the active hostility of its workers, as noted by Warner and Low (1947), Smith (1939) and Gouldner (1954), among others. But in all three of these cases the management was incompetent only in that it took quite insufficient notice of the inevitable human consequences of certain technological changes that were clearly necessary on economic grounds. The most important point in these cases of technological change was not the personal behaviour of the management, but simply who would get what out of the changes. In critical circumstances, then, what matters to the worker (as well as to the other persons who are involved) is simply the division of the spoils. It can hardly be said that the personal behaviour of members of management of any grade would affect the worker's judgement of his task (which is the most important element in his whole work situation) in any circumstances. On the contrary, the finding that the worker's four judgements are independent (cf. chapter III) shows that a transfer from one field of experience to another field of judgement is unlikely. But the contradiction of the recent view that human relations in industry are of prime importance is only apparent. The conclusions stated throughout the argument refer only to the worker's judgements of his situation and not to his behaviour; and his behaviour is certainly not derived simply and directly from his judgements. The point that was made in chapter I is worth repeating, that there must be some relation between the worker's judgements of his situation and his behaviour: if this is not so, one can only suppose that the judgements are made for no practical purpose whatever, which is humanly inconceivable. It can be assumed, therefore, that workers who are satisfied with their tasks will tend to work harder (faster, more accurately, or more efficiently in general) than workers who are not satisfied with their tasks. For the practical purpose of the present argument, the possible consequences in non-productive behaviour (or behaviour in general, within the limits of the work situation) need not be considered. It is sufficient to note that there must be a significant relation between the worker's judgement of his task and his productivity. But this is a particular relation, which does not necessarily hold for the worker's judgements of the other elements of his situation. His judgement of his mates is particularly unlikely to have consequences in productivity, except perhaps in the rather unusual case that the worker's productivity depends directly on the efforts of a working team: that is, the rare case in which the productivity of the group is necessarily and physically determined by the productivity of the slowest or least efficient of its members. The worker's judgement of his pay is possibly more relevant; it is conceivable that the man who is satisfied with his pay might work harder than the man who is not, all other things being equal. The worker's judgement of his bosses, however, is obviously relevant to his behaviour, simply because work (productive behaviour) is what the bosses pay the worker to do. Even in the limiting case that workers can be differentiated solely by the behaviour of their bosses (immediate or higher, or both together), it is likely that the workers whose bosses are more acceptable (in the terms that were defined in chapter VI) would be the more productive. But a situation that is thus limited is not likely. If it is the management's interest to analyse and modify its own behaviour, at whatever level, then it is likely that other elements of the work situation (that is, task conditions and pay) will also be taken into account, simply because the management's responsibility is the whole productive situation. It would follow that the management's initiative may not be as limited as was implied above; and a brief review of certain illustrative cases will suffice to make this clear.

The first that may be taken is that reported by Marriott (1951), in which the worker's effective pay was increased automatically by the reduction of the working week to five days instead of five and a half. The question for the management was whether or not total production could be maintained, by an increased rate of production. Marriott refers the question to two factories that were closely similar in technical respects, and concludes that total production was maintained in one of these, that had a high order of personal leadership on the job (that is, by the immediate boss), and an effective system of joint consultation, and considerable opportunities for promotion; while total production was not maintained (because the rate of production hardly changed) in the other factory in which none of these features were present. It may be, therefore, that production depends on the management's initiative, at least in the case of technological change; but, if it does, it apparently depends on the management's coherent and unitary initiative, and not merely on the applications of this initiative to single isolated aspects of the work situation.

The second case is the reported work of Maier (1952). It is not easy to take this into account because the final report of some years' work is vague in important respects. Firstly, it appears that Maier's work was done chiefly in certain public utilities, rather than in manufacturing industry; and that the work done was concerned chiefly with maintenance tradesmen, the élite of any factory. The aim of the reported programme in applied psychology was to train a number of foremen in efficient methods of handling human relations. The evidence that the foremen's behaviour on the job was changed by training is inconclusive and quite insufficient (as Maier reports it); and Castle (1952) observes that there may be little change in the foreman's attitudes, as he reports them in the test situation, after he has been trained by role playing in human relations. The relevance of Maier's account, however, is not in his debatable views on the training of foremen, but in the implied aim of his work in the factory: which was to give the workers more responsibility for resolving the routine difficulties of their daily tasks. The range of examples given need not be reviewed in detail, for it is clear that in a number of cases various difficulties of production (whether purely technical or personal or both) were solved by the workers themselves. It is also clear that in some cases, at least, the solution could have been arrived at only in this way: in some cases, the same solution would not have been accepted if it had been laid down by the foreman or by his superiors. The importance of Maier's account, then, is that it sets out at length and in detail examples of what may happen in industry without deliberate arrangement or training programmes of the sort which he describes. The point may be made concretely. In both factories C and G a number of men were interviewed who were doing closely similar jobs: storemen, whose task was to move rather heavy stacks of products around the factory on pallet trucks. In factory C, the foreman told his gang in the morning what the day's work was. Who worked with whom, on the tasks that required two or three men, and which person or team went to what location, were determined by the members of the gang among themselves. In factory G, the foremen prescribed exactly who should work with whom, which teams should go to or work from specified locations, and even supervised in person how the teams managed to turn awkward corners and to unload. It cannot be said that there was a difference in productivity in the two situations, because this was not assessed. There was a highly significant difference in the judgements made by the very small groups of workers of their foremen: those in factory C were much more inclined to a favourable opinion. This is not necessarily relevant to the argument, as it may have derived from the very marked differences in general behaviour of the two foremen as persons. What is relevant is the observation that wide differences in supervisory methods may occur in practice, regardless of whether or not the foreman has been trained in the principles of human relations. It can be taken, therefore, that the management's initiative in delegating routine responsibility to the workers can be effective in increasing production, for Maier's evidence that problems were thus solved, that could not have been solved otherwise, is sufficient. The effective range of the management's initiative in this case is obviously sharply limited by the nature of the task, but the general possibility is none the less important.

The third and last case that requires notice is that the management may directly or indirectly increase production by transforming the worker's situation altogether. Jaques (1951) has described the introduction in a factory of an elaborate system of profit-sharing, together with considerable changes in general personnel policy and particular supervisory practices, and with some necessarily involved changes in task conditions. As his account is concerned only with the way in which the scheme was introduced, there is no systematic evidence that it was effective in increasing production. But there is evidence that a number of particular and important problems, which derived from the management's changed general policy, were solved by the introduction of the scheme; and this may be taken as sufficient. It is not necessary to assume that the effectiveness of the scheme depended on the use of the very elaborate methods of joint consultation that are described by Jaques. On the contrary, the detailed and thorough N.I.I.P. report on joint consultation suggests (1952, ch. XVIII) that its usefulness is limited to the solution of minor problems, in ordinary industrial circumstances. What would seem to be important in the case reported by Jaques is the management's initiative in giving the worker a guaranteed status that was far superior to the status of the worker in the general run of factories. The same point may be made by reference to the case reported by Lincoln (1946). In this case, it is clear that what mattered was not the elaborately psychological construction of a new system of human relations but simply the management's guarantee that all workers would have a substantial share both in the firm's development and in its profits; which guarantee has been highly effective for years in increasing production.

The final conclusion, then, is clear: that the worker's behaviour and in particular his productive behaviour, is effectively open to change on the management's initiative. The few findings that have been discussed show that this is so in certain rather unusual cases. The conditions that held in these cases would not necessarily be regarded as appropriate by factory managements generally, for the practical and detailed determination of any management's responsibility to its workers is a complicated derivative of general policy. In any case, policy must depend on the economic and technological possibilities: and it is clear that unusual policies, of the kinds discussed, are possible only under rare conditions, whether within the factory or within whole industries. But, if effective action is possible in unusual circumstances, it is presumably possible in general, if only within rather narrow limits. Within the limits that are set, then, by the nature of the task, the pay system, and the system of supervision, it should be open to any management to take effective action if it wishes.

None the less, it seems likely that a permanent class-war between the managers and the managed (in any sort of industrial society) is to be expected to some extent. The findings that were reported in chapters III to IX are only very indirectly supported by the comparative analyses reported in this chapter. But, in so far as they can be relied upon, they all tend to the same conclusion: that the worker's judgements of his situation are made with reference to his inferior social position, which is accepted as a gross and obvious present fact but not necessarily accepted as being either proper or inevitable.

CHAPTER XI

CONCLUSIONS

THE review of the particular findings has now been taken to the limit and the last task, therefore, is to set out the general conclusions. The aim of the investigation was to explain the worker's judgements of his situation by reconstructing them in the terms of psychological analysis. The reconstructions that have been set out may or may not be credible in general, but they can be accepted only with certain reservations that derive from the nature of the project. These reservations must be considered in detail, but the reader who is prepared to take them for granted can proceed directly to the general conclusions stated in the next section of this chapter, *The worker's general outlook*. The summary of a tediously detailed investigation, however, must take due notice of the limitations of the investigation before the general conclusions can be set out.

The first and most obvious limitation is that of the population which was investigated. It is clear that the seven factories which were studied cannot be regarded as being typical of industry in Melbourne, let alone elsewhere. Thus far, the conclusions stated are necessarily limited. In particular, the sceptical conclusion of chapter X, that management's initiative has a very limited scope in promoting good industrial relations, might well be modified considerably if a study were made of a wider range of factories. However, this is the only substantial reservation that need be made. It is true that the conclusions stated depend on the information given by only 189 men, at the most, or only 127 for the greater part of the analysis. It is also true that these few men are not necessarily representative even of the urban working population from which they were drawn; so that there might be little possibility of generalizing the conclusions put forward to other groups of working men. But this depends simply on the nature of the conclusions. It may be unfortunate that psychology has to rely upon sampling and perhaps, for practical reaons, on distinctly dubious samples. The alternative to the impossible task of complete enumeration is to fall back upon common experience (from which psychology derives, while it extends, corrects and clarifies this experience), and to accept those conclusions that are supported by common experience. The most obvious case is the finding that there is an inverse relation between the worker's judgement of his pay and the number of his dependants. Obvious as this finding is, it could not be accepted without demonstration by particular evidence. But, once demonstrated, it is not unreasonable to expect that it would hold, not only for certain factories in Melbourne in 1952, but also for all sorts of workers anywhere. Of course, it might not hold in detail; the exact number and kind of dependants whom the worker regards as manageable or excessive, in relation to his pay, would doubtless depend on many local circumstances that were outside the scope of the investigation reported. But it is highly unlikely that this finding would be upset in principle by investigations in a wider range.

Much the same point may be made for the conclusions stated about the relation of the worker's judgements of his task, pay and bosses to the factory situation as assessed by the observer. That the worker himself assesses this situation and makes his final judgement partly on the basis of this set of assessments, may be an obvious conclusion, though it too is new in industrial psychology in the detail set out. There is no doubt that investigations in a wider range would modify the conclusions drawn, by extending the classification of desirable and undesirable features in the task, the pay system, or the behaviour of the immediate bosses. But it is, again, inconceivable that the findings stated would be nullified or reversed by a wider range of investigation.

The case with the worker's judgements of his place in society, his expectations for the future and his upbringing is perhaps not so clear because of the complex and tedious derivation of these reconstructed judgements. However, if it is assumed that these judgements can be efficiently defined, the conclusion is again clear. One would expect that any worker's estimation of his place in society would be related to his judgements of his situation at work, although the worker's estimation might be differently defined (or reconstructed) in an investigation with a wider range. The same would hold for the other general judgements reported. In particular, the finding that a happy home life in childhood is related to getting on well with one's workmates is capable of generalization far beyond the limits of the investigation reported.

The second limitation that requires notice derives from the small number of persons studied. The fact that the conclusions stated in the preceding chapters hold (with a few trivial exceptions) regardless of occupational grade, age, length of service or nationality is in a sense misleading. It was necessary to use all the complete interview records that were available, regardless of differences in these four respects, so that differences in the conclusions could hardly have been found. It does not follow at all that there would be no differences in the general case. On the contrary, it is likely that the conclusions stated would not hold indifferently for workers of any occupational grade (tradesmen, process workers or labourers) or of any nationality (by origin). It may be that the differences by grade or national origin would be minor, but it is likely that, in general, they would be significant. It is even more likely that differences by length of service and by age would be not only statistically significant in the general case, but also of psychological importance. In the circumstances of the investigation, it was necessary to draw conclusions indiscriminately for apprentices with a few months' service (at one extreme) and for men with fifty years of experience and up to thirty years of service in the particular firm (at the other extreme). That this wide range of history could not be taken as a particular object of study is a misfortune that must be accepted; but it does not follow that the conclusions stated in the preceding chapters would depend as little on the person's history as they appear to do in the particular circumstances of the investigation.

The third and most important reservation that must be made is that the conclusions drawn depend critically on the method of analysis and, before this, on the general methodological assumptions of the investigators. This set of reservations must be considered at length. It is clear that the limitations of the investigation were in one sense inevitable: a 'natural' experiment (or a field investigation of the sort which the sociologist or anthropologist would regard as appropriate) cannot be conducted with anything like the rigour of the 'true' experiment of the psychological laboratory. It may be unfortunate that the complexity and intractability of the behaviour of ordinary persons in ordinary daily life situations precludes the use of a rigorous experimental design; but the fact must be accepted, and the investigator's only recourse is to the tediously inconclusive comparative analysis which is routine in anthropology and sociology. In this sort of analysis, sound conclusions depend both on the study of a very large number of persons whose situations differ in respects that can be specified exactly and on the inclusion in this study of a wide range of simultaneous information for each individual. The investigation that was undertaken was directed to the second of these requirements, both because it was the easier to meet and because it has been, on the whole, neglected in psychological work. It was not possible to meet the first requirement. The limitations of the investigation are therefore obvious in this respect (as noted above). The emphasis on wide-ranging information had certain less obvious consequences that must be considered in detail.

The analysis made was of the worker's judgements of his situation only, and not of his behaviour at work, for the reasons noted in chapter I. This analysis was of two distinct kinds. The judgements made may be confirmed, extended or clarified in detail by the analysis of the worker's situation as the observer sees it. Thus it appears that the worker may like his task because it is clean, independent or complete, or dislike it because it is the converse; that he may be satisfied with his pay because he can easily get extra payments, or may be dissatisfied because he cannot; and that he may like his immediate bosses because they do not behave in a bossy manner or dislike them because they do. These distinctions depend solely on the observer's analysis of the situation. For the worker's task, this analysis depends on observation of the various tasks. For the worker's pay, it depends on the public records of pay systems, and for the worker's bosses it depends on their own report to the observer of their behaviour as bosses. In all cases, therefore, it is an independent analysis of the worker's situation which in part explains his judgements of the situation. What this analysis does, then, is to show in detail just how the worker assesses his particular situation, by enlarging his reported and not very articulate summary judgements and reconstructing the local grounds of these judgements. The judgements, then, are thus far extended and clarified by the psychologist's analysis of their particular and detailed grounds. Beyond this, there is the analysis of the worker's judgements of his work in relation to certain other judgements of his whole situation: his place in society, his general expectations of other persons (at least in relatively undemanding relations), his perception of his possible future and, lastly, his assessment of his present financial responsibilities. So far as these can be taken to be reliable reconstructions of the worker's relatively inarticulate covert behaviour, they presumably set the limits within which he makes his direct judgements of the work situation; or (to put it another way) they constitute the frame of reference within which he makes these particular judgements. However, these reconstructions too are the psychologist's elaborations of the worker's statements, and are even more remotely derived from the evidence than are the reconstructions of the worker's detailed grounds for judging his work. The reliability or credibility of any of these reconstructions obviously requires consideration, with regard to the analytic procedures that were adopted.

The first step was the content analysis of the interview records, in order to establish a classification of the data. In principle, this was quite a conventional procedure. It is true that the analysis was directed only to the establishment of logically sound categorizations, that were not necessarily metric; and the author has argued elsewhere (1957, ch. X) that this conservative procedure is likelier to give classifications that are psychologically meaningful than is a procedure which imposes metric assumptions on the data from the start. However, the procedure adopted has no distinctive consequences. It is true that what was done is not necessarily something that any psychologist could do, given the same data; but there is no reason why this should be so, as a general condition of experimental work (cf. the author's earlier argument on this, 1957, ch. IV). Further, it is obvious that the present account does not give nearly enough detail for the classifications stated to be directly testable by replication by any other experimenter. But these two limitations are inevitable in any complex and economically reported investigation of the behaviour of persons in real life situations, and in no way depend on the particular method of analysis that was used.

The second step in the analysis was the determination of associations between complex sets of categories, and this did introduce difficulties into the argument. All four of the judgements analysed were treated throughout as simple dichotomies (the worker is satisfied or not satisfied: or more or less satisfied, only the two grades being distinguished). In this, there were no difficulties. The analysis of the detailed grounds of these judgements, from the available evidence of the factory situation itself, raised few difficulties. The worker's assessment of the details of his task, of the pay system under which he works and of the behaviour of his bosses, were all reduced to a final scalar form, the elements of which were all derived from unambiguous dichotomous classifications. The same does not hold for the analysis of the worker's judgements of his general situation. The extreme case is the summary of the worker's view of his place in society (cf. Table 28, in chapter VII). This can be reduced to a minimum of ten categories, by the combination of the men who place themselves definitely in the middle class (as defined in chapter VII) with those who would like to be in the middle class. If this reduction is made, the conclusions stated in chapter VII are not affected; but the final set of ten categories is obviously not a simple scalable classification. It has been argued that the worker's judgement of the appropriateness of his place in society depends on his reference to particular purposes or engagements, so that he makes four somewhat different judgements in relation to his task, pay, bosses and mates. Now this complex set of conclusions can be accepted in detail only on sufficient statistical grounds. In the case taken, there are ten categories that could be written out in a great variety of different orders and which could therefore be divided in a number of different ways in relation to any one of the four judgements that were analysed. For any particular division that was made (and from which conclusions were drawn), it should therefore be possible to say that the categories within either part of the two-fold final classification do not differ statistically, while all the differences between categories in the two separate parts are significant. It must be made quite clear that no such statement can be made. The reported relation between the worker's judgement of his place in society and of his task may be taken as an example. Among ten categories of the former judgement, 45 comparisons can be made of relations with the latter. When these were made, it was found that there were no significant differences among the 6 categories of persons who were finally classified as being satisfied with their tasks, nor among the 4 categories of persons who were dissatisfied with their tasks. But the 24 comparisons of these 6 with the other 4 categories showed significant differences (in conventional terms) in two cases only. It could be argued that the division of the 10 categories that was reported in chapter VII is the best division that could have been made on statistical grounds; and the same point would hold for the divisions that were made in other cases, whether of the same material in relation to other judgements (pay, bosses and mates) or of different but equally complex material (the worker's report of his upbringing and of his general expectations). It must be noted that the obvious mechanical precautions were observed throughout the analysis. All statistical tests depended on the two-tail probability, as the cases in which a specific association or difference (with direction or nature specified) could be proposed were heavily outnumbered by those in which the only reasonable hypothesis was that there should be a significant association or difference. Further, the whole analysis was conducted on alphabetical split-halves of the population, and no findings were accepted as worthy of report that were not statistically significant for both halves. But these routine mechanical safeguards by no means rule out the possibility that the conclusions reported depend only on the spurious accumulation of accidental differences or associations. Whether or not this horrid possibility is likely, it obviously must be considered; and the obvious ground of consideration is the third stage in the analytic procedure, the inference of the psychological meaning of the conclusions drawn.

These conclusions were drawn on dubious grounds, in view of the limitations and difficulties of the first and second stages in the analysis. It is none the less true that the third step supports the first and second: an analysis which is credible and intelligible within the terms of common experience may be preferred to an analysis which is statistically impeccable but not intelligible. The conclusions set out above, then, must be regarded as highly speculative hypotheses rather than firmly demonstrated findings; but they are hypotheses which are intelligible in human terms and which could be exhaustively investigated, given only a sufficient range (a much wider range) of both persons and work situations. The distinction between speculative hypotheses and firm findings is so important that the point must be stressed: the only claim that can certainly be made for the findings reported in the preceding chapters is that they are intelligible and credible but are none the less speculative hypotheses about the grounds of the worker's judgements of his situation. The speculative approach, which depends primarily on a wide range of information, is more likely to give intelligible results than is the approach which depends on the very restricted information that can be gathered by the use of abstract instruments. As the author has argued this point at length elsewhere (1957), it need not be elaborated. The conclusions of the investigation are to be regarded only as speculative hypotheses because the investigation was in all stages directed primarily to the drawing of conclusions that would be psychologically intelligible. So long as this is clear, the way is open for the next stage in the argument, which is the consideration of the relations among the conclusions drawn.

THE WORKER'S GENERAL OUTLOOK

These relations must be considered under two heads. Firstly, there may be relations among the worker's judgements of his situation in general; and this must be considered by itself because of the hypothetical importance of these judgements as the grounds of the worker's behaviour in various situations besides the work situation. Secondly, there is the possibility that the worker's local or particular judgements of his work situation are related to his general judgements of his whole personal situation; so that his final judgements of his situation at work might depend not only on the particular and the general but also on the combinations of the two. This is easily examined by reference to the observed combinations.

The first point may be disposed of briefly. There are no associations at all between the worker's judgement of his place in society, his general expectations of other persons and his general assessment of his future. Whatever dire implication this finding may have for the conventional theory of personality (in terms of general and abstract traits), it is in one sense obvious. The three hypothetical constructs should be independent because they are reconstructions of covert behaviour in three quite distinct modes, or with three quite distinct types of object or purposes. The finding of independence, of course, does not directly validate the elaborate and speculative reconstructions that were made; but it is compatible with the proposition that these reconstructions do correspond to the worker's actual covert behaviour, while a finding of significant relations would hardly have been so compatible. It may be added that one significant relation was noted. The married men who were most co-operative in family affairs (as defined in chapter IX) tended significantly ($P = 0{\cdot}02$) to be those who were brought up happily in ordinary family situations (as defined in chapter VIII). This somewhat obvious finding to some extent adds to the credibility of both its components.

The relations between the worker's final judgements of his work situation and the various combinations of his other judgements (of his work situation in detail, and of his general situation) must be set out at length. The evidence presented is for the whole number of 127 workers (factories A to E). No notice is taken of possible differences by occupational grade, length of service or nationality because, following the argument above, it is not likely that any differences would be found in the particular case even though differences in general might be expected. Differences by age, however, are noted as they were found in this last stage of the analysis. The first set of relations is set out in Table 40, which shows the worker's two judgements of his task conditions in particular (as reviewed in chapter IV) and of his place in society in

TABLE 40

Judgements of the task, its conditions and of place in society

The worker's general judgements are		The task is	
Of the task conditions	Of his place in society	Satisfactory	Not satisfactory
1. Acceptable	Appropriate	33	3
2. Not acceptable	Appropriate	17	11
3. Acceptable	Not appropriate	19	11
4. Not acceptable	Not appropriate	15	18
	Total	84	43

general (as reviewed in chapter VII), against his final judgement of his task. It is clear that the worker is more likely to be satisfied with his task if he is satisfied on both grounds, but the two grounds are independent. The only clear conclusion, therefore, is that the worker is likely to be not satisfied with his task if this is neither suitable in its conditions nor appropriate to his place in society. For this analysis, no differences were found by age.

A similar set of relations is shown in Table 41, the worker's three judgements of the conditions of his pay (the pay system, the nature of his family responsibilities and the appropriateness of his place in society), against his final judgement of his pay. In this case it is obvious that the practical effectiveness of the three general judgements reported can be defined simply by the number of these that are reported as satisfactory. One obvious line of division is between categories 4 and 5 in Table 41: the worker is satisfied with his pay if he is content with any two of the three determinative grounds, the pay system itself, his pay in relation to his present responsibilities or his general place in society. Another equally

TABLE 41

Judgements of pay and of its determinative conditions

The worker's general judgements are			His pay is	
Of the pay system	Of his responsibilities	Of his place in society	Satisfactory	Not satisfactory
1. Satisfactory	Manageable	Appropriate	33	3
2. Satisfactory	Manageable	Not appropriate	8	1
3. Satisfactory	Excessive	Appropriate	19	8
4. Not satisfactory	Manageable	Appropriate	10	3
5. Satisfactory	Excessive	Not appropriate	1	1
6. Not satisfactory	Manageable	Not appropriate	6	4
7. Not satisfactory	Excessive	Appropriate	10	9
8. Not satisfactory	Excessive	Not appropriate	1	10
		Total	88	39

reasonable line of division is between categories 7 and 8: which would give the conclusion that the worker is satisfied with his pay if he is content with any one of the determinative grounds. It must be added that the determinative grounds are not entirely independent. There is a significant tendency ($P < 0{\cdot}01$) for the men who are satisfied with the pay system to regard their place in society as appropriate. If this merely significant relation must be interpreted, the likeliest conclusion is that the worker's judgement of the appropriateness of his place in society may to some extent depend on whether he is well and unconditionally paid for what he does. The concurrence of the two judgements would presumably depend on experience: and the relation noted is highly significant for the middle-aged workers (age 30 or over) but does not hold at all for the young workers (age 29 or less).

The third case to be considered is the worker's judgements of his bosses, which are shown in Table 42 for the 90 men whose immediate bosses were fully interviewed, as noted in chapter VI. There are no differences between these 90 and the remaining 37 in either of the two general judgements referred to, or in the relations of these judgements to the judgement of the bosses as a whole; so that there is no need to set out a reduced version of the table (with two variables only) for the whole group of 127. The first column of the table of course shows (in shorthand form) the classification of immediate bosses that was made from their own reports of their behaviour, as discussed in chapter VI. The worker's classification (or judgement) of all his bosses is shown in the last two columns. In this case too the determinative grounds of the worker's final judgement are little related. There is a just significant association ($P < 0{\cdot}05$) between the first and third grounds only: the men who have good bosses, as defined by the observer from the bosses' own reports, tend somewhat to have no personal expectations beyond that they will go on as they are. As this association is barely significant, it hardly requires interpretation. The comment may be ventured none the less that the older and more experienced men who have found themselves places in which they can work for a good boss are therefore inclined to regard their future simply as a continuation of their present condition. The same general comment can be made as before, that the worker's overall judgement of his bosses may depend on various combinations of the three elements, any of which can over-ride the others. In this case, however, this would be going too far. As in the previous cases, the general statement of relations that is shown in Table 42 could be reduced to a simpler order in a number of ways. If the reduction is made by dividing between categories 1 and 2, the proposition is that satisfaction with bosses in general depends on all three determinative elements being satisfactory, while dissatisfaction depends on all other cases (the number of satisfactory elements in two, one or none). The proposition is significant ($P = 0{\cdot}048$), and holds for 61 of 90 persons, or 68 per cent. The obvious alternative propositions are that the division of Table 42 should be made between categories 4 and 5 or between categories 7 and 8, when it would be supposed that satisfaction with the bosses in general depends on two or one of the determinative elements, respectively, being satisfactory. Both these propositions would give highly significant ($P < 0{\cdot}0001$) statistical associations, but they would hold only for 67 and 59 persons respectively, of 90: or for 74 and 66 per cent respectively. Another possible division of the table is between categories 1, 3, 4 and 7 and categories 2, 5, 6 and 8: or the division simply with regard to the worker's expectations, as the sole element in his overall judgement. This division is again highly significant ($P < 0{\cdot}0001$) and the implied proposition would hold for 71 of 90 persons, or 79 per cent. A possible, though somewhat speculative, conclusion is that the worker's general judgement of his bosses depends firstly or mostly on his having settled down to his situation and accepted the fact that he will never be any other than he now is, a permanent subordinate with a very limited range of responsibilities. The overriding effect of his other judgements may be relatively limited, as is shown in category 2 of Table 42, and even more in categories 5 and 6, as against category 7.

TABLE 42

Judgements of bosses and the conditions of these judgements

	The worker's general judgements are		The bosses are	
The bosses are	Of his place in society	Of his expectations	Satisfactory	Not satisfactory
1. Good	Appropriate	Stay as he is	7	3
2. Good	Appropriate	Hopes to get on	6	5
3. Good	Not appropriate	Stay as he is	8	3
4. Not good	Appropriate	Stay as he is	1	1
5. Good	Not appropriate	Hopes to get on	4	10
6. Not good	Appropriate	Hopes to get on	1	8
7. Not good	Not appropriate	Stay as he is	6	1
8. Not good	Not appropriate	Hopes to get on	—	26
		Total	33	57

The last case to be considered is the worker's judgement of his mates, as set out in Table 43. This is referred only to the worker's general judgements of his place in society and of his upbringing (the latter of which is denoted by a shorthand label that refers to the main distinction between kinds of reported upbringing, as set out in detail in chapter VIII. It may be that the worker's judgement of his mates does depend on his situation; that is, on their behaviour or on the kind of company he keeps. It will be recalled that this possibility could be considered for a few men only (a total of 38) and, at that, only with regard to the limited case of the general comments made on him by the worker's mates. These comments are evidently to a large extent a simple reflection of the worker's behaviour. For the 38 men, there is a highly significant association ($P = 0{\cdot}006$, $\phi = 0{\cdot}51$) between the worker's report of his upbringing (divided as in chapter VIII) and his mates' reports of his

TABLE 43

Judgements of workmates and the conditions of these judgements

The worker's general judgements are		His relations with his mates are	
Of his place in society	Of his upbringing	Satisfactory	Not satisfactory
1. Appropriate	Happy	36	18
2. Appropriate	Not happy	4	9
3. Not appropriate	Happy	20	19
4. Not appropriate	Not happy	3	18
	Total	63	64

acceptability. These latter reports are therefore disregarded. The only reliable evidence of the determinative grounds of the worker's judgement of his mates is that which is shown in Table 43. Here again, the two grounds are independent, so that either may to some extent override the other; but it appears that the judgement of upbringing (or, speculatively, of persons in general) is the more important or more effective. Neither of the simple scalar divisions of Table 43 (categories 1/234 or 123/4) is as effective as the division on upbringing only (categories 13/24). A possible conclusion, therefore, is that the worker's present evaluation of his upbringing may in part determine his assessment of his place in society, in this respect.

SUMMARY

It must be pointed out that the conclusions stated are limited in two respects. It is reasonable to take the particular conclusions of the preceding chapters as being all, in a sense, conclusions about the determinative grounds of the worker's judgements of his situation at work. The worker may judge this situation simply as it is (that is, as it is described by the detached observer) and reach his final conclusions by reference only to his engagements in the work situation. Or he may refer these engagements to his general engagements outside work, as he judges these. Or he may do both at once. In any case, however, it is clear that the final judgements that have been analysed are explained little more, if any more at all, by reference to combined determinative grounds than they are by reference to single grounds. The extent or efficiency of explanation is limited, and a wide margin of unexplained judgements remains.

This may be related to the point made at the beginning of this chapter, that all the conclusions must be regarded as speculative hypotheses rather than firm and definite findings. The limitations of the investigation must be accepted equally with its advantages; but the advantages are not inconsiderable. It may be that any detailed and sustained attempt at the reconstruction of covert behaviour, the behaviour which makes a humanly variable and intelligent link between perception and action, is bound to run into grave practical difficulties merely because this sort of reconstruction has been little attempted in deliberate psychological investigation.

Besides the difficulties that have already been mentioned, there is the obvious point that the information on which the analysis depended could be made greatly more relevant in a further investigation. The definition of the worker's place in society and, even more, the definitions of his general social relations and of his expectations are themselves only speculative hypotheses because the information from which they were derived itself derived from a first and highly provisional attempt to define what was necessary. However, as the reconstruction is necessary in all cases, if only by the inference of the person's intentions on the basis of common experience (on which see the author's argument, 1957, ch. II), it would be as well if it were undertaken deliberately and intelligently. A first step in a largely neglected field, which is difficult to investigate because it is vastly more complicated than is the simple field of experimentally observed stimuli and reactions, is bound to be highly inconclusive in detail but is none the less worth making.

The last point that may be noticed is that the investigation was concerned with the behaviour of relatively unsophisticated persons, of manual workers whose verbal skill is far below that of the professional psychologist and whose dependent and uncreative work situation contrasts sharply with that of the rare workers whose fortune it is to make their own terms of work and their own distinctive achievements. In this case too, it is high time that experimental psychology should drop its exclusive interest in persons in general and should take on some of the interest in persons in particular which is found in social anthropology, sociology or (from a quite different point of view) in clinical psychology. Of course this too means that the first attempts at the exploration of a vast field that has hardly been entered (except in the very limited terms of reference of 'applied' psychology, or the study of industrial efficiency), are also bound to be speculative and inconclusive. They are none the less worth making, and worth reporting.

What has been set out, indeed, is not merely the results of an investigation of industrial behaviour but, within the limits noted, an attempt at a social theory of personality. As a theory, its five distinctive propositions may be stated in general terms.

(i) The person makes considered and coherent judgements of the whole set of his present engagements, both at work (as detailed in chapter III) and outside work. These simple overall judgements may be called summary evaluations.

(ii) The person also makes detailed point by point historical evaluations of his engagements, to a significant extent in the same terms as an observer would use. These detailed judgements are historically prior to the person's summary evaluations but are less general.

(iii) The person also makes historical evaluations of himself in action, in general terms, over a long range of experience, and projecting from the developing range of his experience into his possible future. These general evaluations are also historically prior to his summary judgements, but are more general.

(iv) The summary judgements made derive in part from the person's detailed judgements and in part from his general evaluations. In both cases it is the person's activity in reflecting on his experience and present engagements which makes the summary judgements out of the others. Simply, the evidence is summed up and therefore a conclusion is reached.

(v) But, as all judgements are made in continuing historical situations, there are various possibilities of interaction, (a few simple cases of which were noted in the previous section of this chapter). The general evaluations may be derived from or constructed out of the detailed judgements of present situations. In the long run this must be so: the person makes his general evaluations out of the whole course of his experience. General evaluations, however, may also be reflected back into detailed judgements, and so modify these to some extent. Further, there is the possibility of summary judgements modifying judgements of other kinds. To put it simply: the three kinds of judgements are logically rather than historically distinguishable. They refer to distinct sets of relations: to present engagements as complex wholes, to the details of present engagements and to continuous experience of developing engagements, respectively. Which causes which, in any instantaneous section of the person's developing covert behaviour, may be hard to determine.

It will be noted that the most obvious proposition is omitted:

that the person's summary judgements are the determinative ground of his behaviour. As there is no evidence on this, it must be omitted. It is possible to state, in the most general terms, only a partial attempt at a social theory of personality; but the general statement is worth making, so as to show clearly how far this theory differs from conventional theories of personality.

Even if the theoretical reference is disregarded, the conclusions reported stand. That numerous and, on the whole, firm conclusions could be drawn from an investigation which depended mainly on the unfashionable method of interviewing, is remarkable. It appears that, if one wants to know how a person sees the world he lives in, it is still worthwhile simply to ask him. The factory worker is then easily described. His judgements of his situation at work depend firstly on a realistic appreciation of this situation, in detail. Secondly, his judgements depend on his evaluation of the appropriateness of his place in society, which is complexly derived from his history. Thirdly, his judgements depend on his individual or personal adjustments to both his past (his manner of upbringing) and his future (or expectations). For the worker who is a permanent subordinate, the place of work in life is subordinate. The work that he must do in order to live is tolerable so long as it is acceptable in detail, is not too grossly out of line with his history, and is not totally unpromising for his future.

APPENDIX

THE INTERVIEW CHECK LIST

This is set out exactly as it was used. 'S' at points 34 to 38 means that these were questions to be asked in the words stated and in the order set out.

1. *Identification*

1. Name.
2. Age. 3. Sex. 4. Factory and Department or Section.

2. *Job*

5. Title.
6. Brief description of work-cycle.
7. Jobs before and after in production line.

3. *Perception of Job*

8. Pay: base rate, award rate, overtime, bonuses.
9. Opinion of pay: amount and system.
10. Workmates: who actually work with, how get on with, who else known, how get on with.
11. Who is perceived as boss.
12. Other bosses known.
13. Opinion of each specified boss.
14. How like job: overall opinion.
15. How compare with previous jobs (here and elsewhere).
16. Any alterations in job suggested.
17. Right kind of person for Subject's job.
18. What is a good job generally.

4. *Perception of Factory*

19. Opinion of amenities.
20. What departments known: what do these do.
21. Opinion of factory generally.
22. Any alterations suggested.

5. *Job History*

23. Other jobs: chronological list.
24. Why chosen, why left.
25. Who helped determine: activities and opinions of parents, other relatives, others.
26. At school leaving age: any jobs fancied, other than those tried.
27. (Same period). If you had an open go, what would you have most liked to do when you left school.

28. This factory: all jobs held in order of time.
29. Why chose present job: why chose this factory.
30. Who taught present job, how long to learn.

6. *Union*

31. Member, how long, how active, what has union done in that time.
32. Any other union before.
33. Opinion of unions generally.

7. *Class*

34. S. It is sometimes said that people belong to one class or another. What do you think might be the main classes in Australia?
35. S. To which class do you belong?
36. S. In general, what do you think of the people that belong to the (top, then work downwards) class?
37. S. If all you knew about the following people were their jobs, what class would you say they belonged to?
    1. Clerk.
    2. Executive of a large business—such as the man in charge of all selling for the factory here.
    3. Fitter and turner.
    4. Shop assistant.
    5. Stoker.
    6. Owner of a large business—such as a factory with 200 employees.
    7. Foreman.
    8. Process worker—e.g. unskilled worker.
    9. Owner of small shop—e.g. a suburban newsagent.
    10. Production manager—e.g. the man who runs factory up here.
38. S. What do you think of politics in Australia? (If necessary.) Who do you vote for? Any other? Family vote?

8. *Family of Origin*

39. Father: age, jobs.
40. Mother: age, jobs.
41. Siblings: age, jobs.
42. How brought up: how got on with parents, siblings.
43. Education: where, what standard.

9. *Own Family*

44. Married or single.
45. (If single): why, what about it?
46. (If married): spouse, age, job, education.
47. Children: ages, schools, jobs (if old enough), proposed jobs or schools (if young enough).
48. How bring up children: general statement of methods and aims.
49. How run family: general idea of how decisions are made (to be introduced with examples from important but reasonably 'safe' areas: how decide how to bring up children, how get them into a a job, whether or not to buy something big, who handles the cash).

10. *Leisure Activities*

50. Major activities, as perceived by S. (Prompt on beer and betting.)
51. Friends: how many, what do, how often (if separable from 50). From factory?
52. Detail of use of time: evenings, Saturday, Sunday. (As counterpart of 50, but not down to an accurate and detailed timetable.)

11. *Aspirations*

53. Fears, hopes or intentions for next few years: include how long will stay here.
54. What do when retire.
55. Fantasy: what would do if no need to work at all.
56. Best thing that ever happened to S. (I.e. what he perceives as an outstanding event in his life, without any guidance as to the ground of his judgement.)
57. Worst thing, similarly.
58. What would be the worst thing that could happen to S?
59. Best thing?
60. Why work? (If not self-evident, as with Ss. who have dependants, or have at least to keep themselves, *ask.*)
61. What did you think of the interview?
62. I's prognosis, or general evaluation of S.

REFERENCES

ARGYLE, M., GARDNER, G., and CIOFFI, F. 1956. 'The measurement of supervisory methods'. Unpublished report to the *Department of Scientific and Industrial Research and the Medical Research Council* (Great Britain). London, pp. 1–49.

BAKKE, E. W. 1933. *The Unemployed Man.* London. Nisbet.

BARKIN, S. 1950. 'A trade unionist appraises management personnel philosophy'. *Harvard Business Review*, 28, pp. 59–64.

CASTLE, P. F. C. 1952. 'The evaluation of human relations training for supervisors'. *Occup. Psychol.*, 26, pp. 191–205.

CENTERS, R. 1948. 'Occupational mobility of urban occupational strata'. *Amer. Sociol. Rev.*, 13, pp. 197–203.

CENTERS, R. 1949. *The Psychology of Social Classes.* Princeton, New Jersey. Princeton University Press.

CHRISTENSEN, E. H. 1953. 'Physiological valuation of work in Nykroppa iron works'. Chapter 10 in Floyd, W. F., and Welford, A.T. (Eds.), *Symposium on Fatigue.* London. H. K. Lewis.

GARDNER, B. B., and MOORE, D. G. 1950. *Human Relations in Industry.* Chicago. Irwin.

GOULDNER, A. W. 1954. *Wildcat Strike.* Yellow Springs, Ohio. The Antioch Press. London. Routledge and Kegan Paul.

HERON, A. 1954. 'Satisfaction and satisfactoriness: complementary aspects of occupational adjustment'. *Occup. Psychol.*, 28, pp. 140–53.

JACKSON, J. M. 1953. 'The effect of changing the leadership of small work groups'. *Hum. Rel.*, 6, pp. 25–44.

JAQUES, E. 1951. *The Changing Culture of a Factory.* London. Tavistock Publications.

JONES, E. 1920. *Papers on Psycho-Analysis.* London. Baillière, Tindal and Cox.

KATZ, D., and KAHN, R. L. 1952. 'Some recent findings in human-relations research in industry'. Pp. 650–65 in Swanson, G. E., Newcomb, T. M., and Hartley, E. L. (Eds.), *Readings in Social Psychology.* New York. Henry Holt.

KRECH, D., and CRUTCHFIELD, R. S. 1948. *Theory and Problems of Social Psychology.* New York. McGraw-Hill.

LAFITTE, P. 1957. *The Person in Psychology: Reality or Abstraction.* London. Routledge and Kegan Paul.

LINCOLN, J. F. 1946. *Lincoln's Incentive System.* New York. McGraw-Hill.

MACE, C. A. 1948. 'Satisfactions in work'. *Occup. Psychol.*, 22, pp. 5–12.

MACIVER, R. M. 1942. *Social Causation.* Boston. Ginn.

MAIER, N. R. F. 1952. *Principles of Human Relations.* New York. Wiley.

MARRIOTT, R. 1951. 'Socio-psychological factors in productivity'. *Occup. Psychol.*, 25, pp. 15–24.

MILLER, D. C., and FORM, W. H. 1951. *Industrial Sociology.* New York. Harper.

NATIONAL INSTITUTE OF INDUSTRIAL PSYCHOLOGY. 1952. *Joint Consultation in British Industry.* London. Staples Press.

OESER, O. A., and HAMMOND, S. B. (Eds.). 1954. *Social Structure and Personality in a City.* London. Routledge and Kegan Paul.

REFERENCES

PILGRIM TRUST. 1938. *Men Without Work*. Cambridge University Press.

REYNOLDS, L. G., and SHISTER, J. 1949. *Job Horizons*. New York. Harper.

RICE, A. K., HILL, J. M. M., and TRIST, E. L. 1950. 'The representation of labour turnover as a social process'. *Hum. Rel.*, 3, pp. 349–72.

ROETHLISBERGER, F. J., and DICKSON, W. J. 1939. *Management and the Worker*. Cambridge, Mass. Harvard University Press.

RUNDQUIST, E. A., and SLETTO, R. F. 1936. *Personality in the Depression*. Minneapolis. University of Minneapolis Press.

SCOTT, W. 1950. *Greater Production*. Sydney. Law Book Co. of Australia.

SMITH, E. D. 1939. *Technology and Labour*. New Haven. Yale University Press.

TRIST, E. L., and BAMFORTH, K. W. 1951. 'Some social and psychological consequences of the long-wall method of coal getting'. *Hum. Rel.*, 4, pp. 3–38.

VITELES, M. S. 1954. *Motivation and Morale in Industry*. New York. Norton. London. Staples.

WALKER, J., and MARRIOTT, R. 1951. 'A study of some attitudes to factory work'. *Occup. Psychol.*, 25, pp. 181–91.

WARNER, W. L., and LOW, J. O. 1947. *The Social System of the Modern Factory*. New Haven. Yale University Press.

WREN, H. A. 1942. *Vocational Aspiration Levels of Adults*. New York. Bureau of Publication, Teachers' College, Columbia University.

INDEX